FLORA ZAMBESIACA

Flora terrarum Zambesii aquis conjunctarum

T0141406

VOLUME THIRTEEN: PART TWO

FLORA ZAMBESIACA

MOZAMBIQUE

MALAWI, ZAMBIA, ZIMBABWE

BOTSWANA

VOLUME THIRTEEN: PART TWO

Edited by
J.R. TIMBERLAKE & E.S. MARTINS

on behalf of the Editorial Board:

D.J. MABBERLEY
Royal Botanic Gardens, Kew

M.A. DINIZ
*Centro de Botânica, Instituto de Investigação
Científica Tropical, Lisboa*

J.R. TIMBERLAKE
Royal Botanic Gardens, Kew

Published by the Royal Botanic Gardens, Kew,
for the Flora Zambesiaca Managing Committee
2010

PLANTS PEOPLE
POSSIBILITIES

First published in 2010 by
Royal Botanic Gardens, Kew
Richmond, Surrey, TW9 3AB, UK
www.org.uk

ISBN 978 1 84246 209 6

British Library Cataloguing in Publication Data
A catalogue record for this book is available from the British Library

Typesetting by Christine Beard
Publishing, Design & Photography
Royal Botanic Gardens, Kew

Printed in the UK by Marstons Book Services Ltd

For information on or to purchase all Kew titles please visit
www.kewbooks.com or e-mail publishing@kew.org

All proceeds go to support Kew's work in saving the world's plants for life

CONTENTS

FAMILIES INCLUDED IN VOLUME 13, PART 2

ERIOSPERMACEAE

DRACAENACEAE

ARECACEAE (PALMAE)

PONTEDERIACEAE

BROMELIACEAE

MAYACACEAE

No new names or combinations published in this part.

ERIOSPERMACEAE

by P.L. Perry

Geophytes with leaves produced later than or with the flowers. Tuber hypocotyledonaly simple or multiple, sometimes with a proliferation of narrow rhizome-like outgrowths, globose or irregular in shape, internal flesh white, yellow, pink or red; with one or more apical or basal growing points. Leaves 1–several, each with a petiole-like leaf sheath forming a fibrous or membranous neck with age; lamina linear to ovate or orbicular, erect or prostrate, glabrous or pubescent. Inflorescence a few- to many-flowered simple raceme; peduncle base sheathed by a bract which may remain below ground or appear green and leaf-like above ground. Bracts minute; pedicels very short to long. Flowers bisexual, diurnal, bell-shaped or wheel-shaped, sometimes scented. Tepals 6, shortly united at base, equal or sub-equal, white, very pale-green or yellow with dark median nerves. Stamens 6, attached to base of tepals; filaments equal to sub-equal, filiform or lanceolate, apiculate; anthers dorsifixed, introrse. Ovary superior, trilocular, globose to ovoid; style terete; stigma small, papillose; ovules few, axile, in 2 rows on placenta. Capsule obovoid, emarginate, loculicidally dehiscent. Seeds comma-shaped, covered with white, 1-celled hairs, turning brown with age.

One genus, *Eriospermum*, with more than 100 species confined to sub-Saharan Africa, concentrated in the Western Cape of South Africa.

The family Eriospermaceae has been sunk into Ruscaceae/Convallariaceae by Rudall, Conran & Chase (Bot. J. Linn. Soc. **134**: 73–92, 2000), but is here kept separate.

ERIOSPERMUM Willd.

Eriospermum Willd., Sp. Pl., ed.4 **2**(1): 110 (1799). —Perry in Contrib. Bolus Herb. **17**: 1–320 (1994).

Characters as for the family.

1. Plants with single leaf only, produced after the flowers 7
 – Plants with 2–several leaves, produced with the flowers 2
2. Tepals yellow, mid-nerve green 4
 – Tepals white to pinkish, mid-nerve blue to pink 3
3. Flower urn-shaped, tepals becoming recurved; leaves erect with contemporary leaf sheath exserted up to 50 mm . . . . **5.** *kirkii*
 – Flowers bell-shaped, tepals spreading; leaves erect but contemporary leaf sheath barely exserted . . . . **6.** *bakerianum*
4. Leaves 2–5, ovate, large, up to 90 × 40 mm, glabrous to pubescent; inflorescence many-flowered and regularly-shaped . . . . **1.** *mackenii*
 – Leaves 1–3, not ovate and large, glabrous or basally hairy; inflorescence few- to several-flowered, regularly or irregularly-shaped or sub-corymbose 5
5. Leaves 3, glabrous; leaf sheath obvious, mottled towards base; inflorescence sub-erect, lax, regularly-shaped . . . . **2.** *triphyllum*
 – Leaves 1–2, glabrous or basally hairy, leaf sheath not exserted, or if exserted not mottled; infloresence sub-corymbose to irregularly-shaped 6
6. Leaf sheath exserted up to 30 mm above ground; infloresence sub-corymbose; limited to E Zimbabwe and adjacent Mozambique . . . . **3.** *cecilii*

- Leaf sheath not exserted, or by less than 15 mm; inflorescence few-flowered, irregularly-shaped; widespread from Zambia to Botswana and N South Africa . . . . **4.** *porphyrovalve*
7. Inflorescence with large peduncular bract similar to foliage leaf 8
- Inflorescence with a peduncular bract considerably smaller than foliage leaf . . 9
8. Tuber flesh bright yellow; inflorescence a many-flowered compact raceme . . . . **7.** *roseum*
- Tuber flesh white; inflorescence few- to several-flowered, corymbose . . . . **8.** *rautanenii*
9. Flowers bell-shaped; tepals yellow; filaments filiform . . . . **9.** *flagelliforme*
- Flowers wheel-shaped; tepals green; filaments lanceolate . . . . **10.** *kiboense*

1. **Eriospermum mackenii** (Hook. f.) Baker in J. Linn. Soc., Bot. **15**: 266 (1876); in Fl. Cap. **6**: 378 (1896). —Whitehouse in F.T.E.A., Eriospermaceae: 2 (1996). Type: South Africa, KwaZulu-Natal, cultivated at Kew, 21.vi.1874, *Macken* s.n. (K holotype).

   Bulbine mackenii Hook. f. in Bot. Mag. **28**: t.5955 (1872).

Plants with flowers and leaves together; solitary inflorescence up to 40 cm high. Tuber simple, globose to sub-globose, sometimes becoming compound and irregular with age, up to 40 × 45 mm; skin dark brown, rough, internal flesh pink to red; growing point apical. Old leaf sheaths forming a fibrous neck up to 65 × 25 mm. Leaves 2–5, erect to sub-erect; contemporary leaf sheaths barely exserted; lamina ovate, basally chanelled, up to 90 × 40 mm, glabrous to pubescent. Peduncle up to 30 cm long and 3 mm in diameter, glabrous to basally pilose. Raceme up to 200 × 35 mm, with up to 50 flowers. Bracts ovate, 1–2 mm long. Pedicels to 25 mm long. Flowers to 12 mm in diameter, bell-shaped, becoming recurved. Tepals equal, strap-shaped, 9–10 × 2 mm, bright yellow with a green mid-nerve. Filaments to 6 mm long, filiform, yellow. Ovary 2.5 × 2 mm, ovoid, pale green; style cylindrical, 4.5 mm long, white. Capsule up to 8 × 7 mm, obovoid.

Three subspecies in the Flora area, differentiated by details of pubescence.

1. Plants entirely glabrous in all parts . . . . subsp. *mackenii*
- Plants partially pubescent 2
2. Contemporary leaf sheath, base of lamina and base of peduncle covered in short hairs . . . . subsp. *galpinii*
- Contemporary leaf sheath, entire lamina and base of peduncle covered in short or longer hairs . . . . subsp. *phippsii*

Subsp. **mackenii**. —Perry in Contrib. Bolus Herb. **17**: 34 (1994). FIGURE 13.2.1.

   Eriospermum junodi Baker in Bull. Herb. Boiss., sér.2 **1**: 783 (1901). Type: South Africa, KwaZulu-Natal, Pinetown, *Junod* 149 (Z holotype).

   Eriospermum veratriforme Poelln. in Feddes Repert. Spec. Nov. Regni Veg. **52**: 121 (1943). Type: Tanzania, Dodoma Dist., Uyansi, between Chaya and Tura, 5.i.1926, *Peter* 34329 (B holotype).

Zambia. B: Mongu Dist., 61 km from Mongu to Kaoma, fr. 30.i.1975, *Brummitt, Chisumpa & Polhill* 14185 (K, SRGH). N: Kaputa Dist., Lake Mweru-Wantipa, 17.xii.1960, *Richards* 13745 (SRGH). S: Kazungula Dist., Machile, 4.xii.1960, *Fanshawe* 5939 (SRGH). **Zimbabwe**. W: Bulawayo, Lochview, fl. 13.xii.1975, *Cross* 332 (SRGH). C: Gweru (Gwelo), Senka Township, fl. 21.xi.1965, *Biegel* 581 (SRGH). E: Mutare Dist., Odzi Road, Save (Sabi) Drift, fl. 1.xii.1954, *Wild* 4665 (BR, M, SRGH).

Mozambique. MS: Manica Dist., near Chimoio, Garuso (Garuzo), fl. 29.xi.1943, *Torre* 6254 (LISC) & fl. 14.x.1943, *Torre* 6298 (LISC). M: Namaacha, Mt Ponduine, c.600 m, fl. 7.x.1978, *de Koning* 7306 (LMU).

Also in Tanzania and N & E South Africa. Rocky grassland; 50–1400 m.

Conservation notes: Widespread taxon; not threatened.

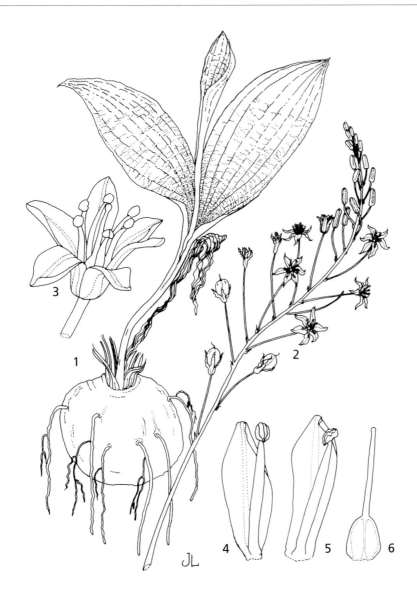

Fig. 13.2.1. ERIOSPERMUM MACKENII subsp. MACKENII. 1, plant with tuber and leaves (× ²/₃); 2, inflorescence (× ²/₃); 3, flower (× 2); 4, outer tepal and stamen (× 3); 5, inner tepal and stamen (× 3); 6, gynoecium (× 3). All from *Perry* 1698. Drawn by Jeanette Loedolff. From Contrib. Bolus Herb. (1994)

Subsp. **galpinii** (Schinz) P.L. Perry in Contrib. Bolus Herb. **17**: 34 (1994). Type: South Africa, Mpumulanga, Barberton, banks of Queens' R., 1890, *Galpin* 1135 (Z lectotype, BOL, K, PRE).

Eriospermum galpinii Schinz in Bull. Herb. Boiss. **4**: 416 (1896). —Baker in Fl. Cap. **6**: 378 (1896).

Eriospermum reflexum Schinz in Bull. Herb. Boiss. **2**(1): 858 (1901). Type: Namibia, Hereroland, between Otjihenene & Seis, *Dinter* 1337 (Z).

Eriospermum omahekense Engl. & Krause in Bot. Jahrb. **45**: 184 (1923). Type: Namibia, near Omaheke, xii.1908, *Dinter* 678a (not located).

Eriospermum pilosopetiolatum Dinter in Feddes Repert. **19**: 184 (1923). Type: Namibia, Auasberge, Farm Lichtenstein, i.1923, *Dinter* 4352 (B).

Eriospermum convallariaefolium Dinter in Feddes Repert. **30**: 83 (1932). Type: Namibia, S foot of Auasberge, 28.ii.1923, *Dinter* 4534 (B).

Botswana. N: Okavango, one island of Nqautsha group, 20.xi.1979, *P.A. Smith* 2889 (SRGH). SW: Ghanzi Dist., c.24 km W of Ghanzi, 1070 m, fl. 20.x.1969, *Brown* 6738 (K, SRGH). **Zimbabwe**. C: Gweru Dist., Watershed Block, by termitaria, 1400 m, fl. 23.xi.1967, *Biegel* 2330 (K, SRGH). **Mozambique**. M: Namaacha Dist., Goba Fronteira, fr. 11.i.1980, *de Koning* 7970 (LMU).

Also in South Africa and N Namibia. Mixed savanna or grassland; 400–1400 m.

Conservation notes: Widespread taxon, but disjunct distribution; not threatened.

Subsp. **phippsii** (Wild) P.L. Perry. in Contrib. Bolus Herb. **17**: 39 (1994). Type: Zimbabwe, Chimanimani Mts, Musapa Gap, 20.xii.1957, *Phipps* 840 (SRGH holotype, K).

Eriospermum phippsii Wild in Kirkia **4**: 134 (1964).

Zimbabwe. E: Chimanimani Mts, 1670 m, fl. 14.i.1974, *Bamps, Symoens & Vanden Berghen* 839 (BR, SRGH); Chimanimani Dist., Chikukwa's Kraal near Martin Forest Reserve, fl. 14.xi.1967, *Mavi* 613 (K, SRGH).

Not known elsewhere. Quartzite grassland above the level of *Brachystegia tamarinoides* vegetation; 1400–1800 m.

Conservation notes: Endemic to the Chimanimani Mts in Zimbabwe, but probably not under threat; Lower Risk near threatened.

2. **Eriospermum triphyllum** Baker in Bot. Jahrb. Syst. **15**: 471 (1892); in F.T.A. **7**: 473 (1898). —Whitehouse in F.T.E.A., Eriospermaceae: 3 (1996). Type: Kenya, Machakos/Kitui Dist., Ukamba, 1877, *Hildebrandt* 2651 (K holotype, BM).

Eriospermum linearifolium Baker in F.T.A. **7**: 473 (1898). Type: Kenya, Teita Dist., Mbuyuni, 1893, *Scott-Elliot* 6203 (K holotype, BM).

Plants with flowers and leaves together; inflorescence up to 20 cm high. Tuber globose, up to 22 × 18 mm; skin rough, pale fawn, internal flesh pink; growing point apical, several on older tubers. Old leaf sheaths up to 30 mm long and 2 mm in diameter, softly fibrous. Leaves erect, 1–3 per growing point; contemporary leaf sheath exserted up to 15 mm, mottled with maroon towards base; lamina up to 120 × 10 mm, linear, narrowly lanceolate or falcate, bright-green, glabrous. Peduncle to 95 mm long, less than 1 mm wide, basal 10 mm maroon-spotted, green above. Raceme sub-erect, lax, to 80 × 30 mm, with c.10 flowers. Bracts 1–1.5 mm long, membranous. Pedicels to 21 mm long, spreading-erect. Flowers spreading, to 16 mm in diameter. Tepals sub-equal, linear, yellow with red stippling on underside; outer to 8 × 2 mm, inner slightly broader and shorter often with hooded apex. Filaments erect, filiform, 5 × 0.5 mm, yellow. Ovary ovoid, 3-lobed, 2 × 1.75 mm; style narrowly cylindrical, 4 mm long, 0.5 mm in diameter, yellow.

Zambia. C: Kabwe Dist., between Chipepo and Lukanga R., 60 km NW of Kabwe, 1140 m, fl. 20.i.1973, *Kornaś* 3058 (K). S: Kazungula Dist., Machile, fl. 27.xii.1960,

Fanshawe 6027 (K). **Zimbabwe**. W: Bulawayo, Luveve Teacher's College, 1350 m, fl. 30.xi.1960, *Norman* R58 (K). S: Chibi Dist., c.8 km E of Chivi Village, fl. 7.xii.1961, *Leach* 11332 (K, SRGH). **Mozambique**. MS: Chimoio Dist., Belas cliffs, fl. 26.ii.1948, *Garcia* 395 (LISC).

Also in Kenya, Tanzania, Uganda and Ethiopia. Rocky soil in grassland and damp depressions (F.T.E.A.); 700–1800 m.

Conservation notes: Widespread species; not threatened.

3. **Eriospermum cecilii** Baker in Bull. Misc. Inform., Kew **1906**: 28 (1906). Type: Zimbabwe, Nyanga Mts, xii.1899, *Cecil* 204 (K holotype).

Plants with leaves and flowers together; inflorescence up to 120 mm high. Tuber not known. Old leaf sheaths to 65 × 13 mm, softly fibrous. Leaves 1–2, erect; contemporary leaf sheaths exserted to 20 mm, terete, 1 mm in diameter; lamina narrowly elliptic to lanceolate, to 47 × 12 mm, subcoriaceous, glabrous. Peduncle to 80 mm long, 1.5 mm in diameter, glabrous. Raceme subcorymbose, 45 × 40 mm, with 8–16 flowers. Bracts boat-shaped, 1.5–2 mm long. Pedicels spreading-erect, to 20 mm long. Flowers bell-shaped. Tepals linear-oblong, 6–7 × 1.5–2 mm, yellow. Filaments filiform, yellow; anthers oblong, small, yellow. Ovary globose; style longer than ovary.

Zimbabwe. E: Nyanga Dist., Nyanga Nat. Park, Mare R., fl. 19.ii.1947, *Wild* 795 (K, SRGH); 4 km E of Rugare, c.2300 m, fl. 22.xii.1973, *Bamps, Symoens & Vanden Berghen* 168 (BR, SRGH).

Not known elsewhere. Growing among wet rocks and in open grassland; 1800–2300 m.

Conservation notes: Apparently endemic to the Nyanga Mts; possibly Vulnerable. The Sabonet Red Data List (Golding 2002) states it is Data Deficient in Zimbabwe and possibly only near-endemic.

4. **Eriospermum porphyrovalve** Baker in J. Bot. **29**: 71 (1891); in Fl. Cap. **6**: 371 (1896). Type: South Africa, former Transvaal, Houtbosch, 1883, *Rehmann* 5765 (K holotype, Z).

  Eriospermum tenellum Baker in Fl. Cap. **6**: 377 (1896). Types: Botswana, Bakwena territory, v.1883, *Holub* s.n. (K syntype) & South Africa, Pretoria, Wonderboompoort, 1883, *Rehmann* 4468 (K syntype, BM, BOL, Z).

  Eriospermum dissitiflorum Baker in Bull. Herb. Boiss., ser.2 **4**: 997 (1904). Type: South Africa, Magaliesberg, 4.xi.1893, *Schlechter* 3622 (GRA, K, PRE, Z).

Plants with flowers and leaves together; inflorescence up to 150 mm high. Tuber small, globose, becoming irregular in shape and multiple with age, 10–15 × 1–18 mm; skin light brown, very thin, internal flesh white to pinkish; growing point apical, up to 10 shoots on one tuber. Old leaf sheaths to 50 × 5 mm, coarsely fibrous. Leaves normally 2 per growing point, erect or suberect; leaf sheaths not exserted, or to 15 mm long; lamina linear, c.60 × 4 mm or elliptic, 10 × 5 mm, green, glabrous both surfaces or basally hairy. Peduncle to 100 mm long. Racemes 2–3 per growing point, irregularly-shaped, lax, 45 × 40 mm, with 8–10 flowers. Bracts ovate, 1 mm long. Pedicels to 35 mm long, arcuate. Flowers bell-shaped, becoming recurved, to 12 mm in diameter. Tepals equal to subequal, 5 × 2 mm, bright yellow with a green mid-nerve, apically red streaked. Filaments filiform, 4 mm long, yellow. Ovary globose to ovoid, 1.75–2 × 1.5 mm, green; style 3 mm long, yellow.

Botswana. N: Kwando R., near Tsimanemcha, 28.i.1978, *P.A. Smith* 2298 (BOT, PRE PSUB). SE: Gaborone, Content Farm, 1050 m, fl. 12.x.1977, *Hansen* 3221 (BOT, BR, K, PRE, SRGH). **Zambia**. W: Mwinilunga Dist., 6 km N of Kalene Hill, fl. 12.xii.1963, *Robinson* 5903 (K, SRGH). S: Kazungula Dist., Machile, fl. 24.xi.1960, *Fanshawe* 5917 (SRGH). **Zimbabwe**. N: Gokwe Dist., Sengwa Research Station, fl. 6.ii.1968, *Jacobsen*

100 (SRGH). W: Matobo Dist., Farm Besna Kobila, 1450 m, fl. i.1953, *Miller* 1499 (K, LISC, SRGH). C: Gweru Dist., 28 km SSE of Kwekwe, c.1300 m, fl.& fr. 6.ii.1966, *Biegel* 916 (K, SRGH). E: Mutare Dist., Tsungwezi R., c.900 m, fl.& fr. 30.xi.1954, *Wild* 4652 (BR, M, SRGH). S: Masvingo Dist., Lake Mutilikwe (Kyle Dam) Game Park, fl. 10.xi.1969, *Kelly* 132 (SRGH).

Also in South Africa. Seasonally wet or marshy areas in annual grassland and shallow sandy granite soils; 900–1500 m.

Conservation notes: Widespread species; not threatened.

5. **Eriospermum kirkii** Baker in J. Linn. Soc., Bot. **15**: 267 (1876); in F.T.A. **7**: 474 (1898). Type: Mozambique, Lower Zambezi R., between Lupata & Sena, 1858, *Kirk* s.n. (K holotype). FIGURE 13.2.**2**.

Plants flowering with leaves; inflorescence up to 250 mm high. Tuber globose to sub-globose, about 3 × 28 mm or 20 × 45 mm; skin rough, darkish brown, internal flesh red; growing point apical, 1–several per tuber. Old leaf sheaths to 65 × 20 mm, chestnut-brown, membranous. Leaves erect, 1–2 per growing point; contemporary leaf sheath exserted to 50 mm, 1.5–2 mm wide, straight or curved, dark red becoming lighter towards lamina, glabrous; lamina elliptic to lanceolate, to 100 × 38 mm, mucronate, bright green, somewhat coriaceous, glabrous with prominent parallel veins. Peduncle to 200 × 2 mm basal 20 mm mottled maroon, bright green above. Racemes 1–2 per growing point, compact to lax, 55–80 × 20–50 mm, with 12–15 flowers. Bracts triangular, 2–3.5 mm long, straight or recurved, brown, membranous. Pedicels to 20 mm long, spreading-erect, mottled red. Flowers urn-shaped, c.10 × 12 mm. Tepals recurved when fully open, equal or sub-equal, united for 2 mm, ligulate, acute, white to pale pink with a broad red-streaked mid-nerve; outer ones 8–9 × 2 mm, inner c.10 × 2 mm. Filaments joined to base of tepals for 1–2 mm, filiform, apiculate, to 2 mm long, white. Ovary ovoid, c.2 × 1.5 mm, green; style cylindrical, nearly 2 mm long, white. Capsule ovoid, emarginate, 6 mm long × 8 mm wide; tepals persistent.

Zambia. E: Chadiza Dist., between Chadiza turn-off and Nsadzu R., 900 m, fl. 27.xi.1958, *Robson* 741 (BR, K, LISC, SRGH). **Malawi**. N: Karonga Dist., 26 km W of Karonga, 730 m, fl.& fr. 31.xii.1976, *Pawek* 12128 (BR, K, MAL, MO, SRGH). S: Machinga Dist., Balaka, Toleza Expt. Farm, fl.& fr. 4.xii.1956, *Jackson* 2086 (BR, K, MAL, SRGH); Mangochi (Fort Johnston), Kadewere village, E of Lake Malombe, 14.ix.1963, *Salubeni* 102 (MAL, SRGH). **Mozambique**. Z: Lugela Dist., Namagoa plantations, fl.& fr. ix.1945, *Faulkner* 194 (BR, K, SRGH); Alto Molócuè Dist., 13 km from Alto Molócuè to Gilé, 500 m, fl.& fr. i.xii.1967, *Torre & Correia* 15299 (LISC).

Not known elsewhere. Poor dry, sandy soils on grassy flats or open mopane or miombo woodland; 50–1000 m.

Conservation notes: Endemic to eastern parts the Flora Zambesiaca area. A widespread species; not threatened.

6. **Eriospermum bakerianum** Schinz in Verh. Bot. Vereius Prov. Brandenburg **31**: 215 (1890). —Baker in F.T.A. **7**: 471 (1897). —Sölch *et al.* in Merxmüller, Prodr. Fl. SW Afr., fam. 147: 46 (1970) as *"bakeranum"*. Type: Namibia, Amboland, Olukondo, 2.xii.1885, *Schinz* 25 (K holotype).

Plants with leaves and flowers together; inflorescence up to 250 mm high. Tuber small to medium sized, globose, becoming elongated vertically or multiple with age, to 35 × 65 mm; skin light- to greyish-brown, tough and wrinkled, internal flesh dark maroon-red; growing point apical, 1–several per tuber. Old leaf sheath forming neck of fine fibres to 20 × 100 mm, light chestnut-brown. Leaves 1–4 per growing point, erect; contemporary leaf sheath barely exserted; lamina narrowly lanceolate to elliptic, sheathing at base, to 120 × 11 mm or 55 × 28 mm, somewhat fleshy to leathery, green on both surfaces, glabrous. Peduncle 60–80 mm long, green, glabrous. Racemes 1–4 per growing point, lax, to 120 × 70 mm, with up to 16 flowers. Bracts triangular, 2 mm long, membranous. Pedicels up to 70 mm long, green, glabrous. Flowers bell-shaped becoming recurved when fully open. Tepals sub-equal, narrowly elliptic, white with a dark reddish-brown mid-nerve, outer to 9 × 2 mm, inner to 11 × 3 mm. Filaments joined to base of tepals for c.2 mm, sub-equal, narrowly lanceolate, outer c.6 mm long, inner 7 mm long. Ovary ovoid, c.3 × 2.5 mm, white; style narrowly cylindrical, 3.5–4 mm long, white. Capsule to 7 mm long × 6 mm wide. Seeds 3 mm long, with hairs 5–6 mm long.

Fig. 13.2.**2**. ERIOSPERMUM KIRKII. 1, plant with tuber and leaves (× ²/₃); 2, inflorescence (× ²/₃); 3, flower (× 3); 4, outer tepal and stamen (× 6); 5, inner tepal and stamen (× 6); 6, gynoecium (× 6). All from *Goldblatt* 7522. Drawn by Jeanette Loedolff. From Contrib. Bolus Herb. (1994).

Subsp. **bakerianum**. —Perry in Contr. Bolus Herb. **17**: 50 (1994).

> *Eriospermum bechuanicum* Baker in F.T.A. **7**: 472 (1898). Type: Botswana, Ngamiland, near Kwebe, xii.1896, *E. Lugard* 80 (K holotype).
> *Eriospermum lineare* Poelln. in Feddes Repert. **53**: 183 (1944). Type: Namibia, Hereroland, Aitsas, 19.xii.1908, *Dinter* 813 (SAM).

Botswana. N: Ngamiland, near Maun to Mababe road, *P.A. Smith* 4350 (K, MO, NBG, PRE, S). SW: Ghanzi commonage, c.950 m, fl. 10.i.1970, *Brown* 7569 (K, SRGH). SE: Mahalapye, fl. iii.1914, *Rogers* 6300 (SRGH).

Also in Namibia and N South Africa. Flat areas of coarse sandy loam soil in open or overgrazed areas; 900–1100 m.

Conservation notes: A moderately widespread species, although in Botswana possibly confined to Kalahari sands; not threatened.

The other subspecies, subsp. *tortuosum* (Dammer) P.L. Perry, is a smaller, more delicate plant, locally common in C Namibia.

7. **Eriospermum roseum** Schinz in Bull. Herb. Boiss. **4**, App.3: 38 (1896). —Baker in F.T.A. **7**: 474 (1898). Type: Namibia, Rehoboth, *Fleck* 888 (Z holotype).

> *Eriospermum majanthemifolium* Krause & Dinter in Bot. Jahrb. **45**: 141 (1910). Type: Namibia, Okahandja, 9.i.1907, *Dinter* 389 (SAM).

Plants with leaves appearing after the flowers; inflorescence up to 220 mm high arising with a leaf-like peduncular bract. Tuber solitary or multiple, globose, becoming elongated vertically with age, to 70 mm from apex to base and 40 mm wide; skin light to dark brown, leathery, internal flesh bright yellow; growing point apical, one per tuber. Old leaf sheaths to 80 × 10 mm, greyish-brown, membranous. Leaf solitary, spreading-erect; contemporary leaf sheath exserted approximately 15 mm, 2.5 mm wide, with groove on adaxial side, whitish, densely covered with short white hairs; lamina orbicular-cordate, to 110 × 134 mm wide, green on both surfaces, glabrous or sparsely hairy on lower side near base. Peduncular bract leaf-like, sub-erect, broadly orbicular-cordate, up to 500 × 85 mm, green with silvery sheen on underside, lower 8 mm encircling peduncle, white, finely pilose. Peduncle to 170 × 1.5–2 mm, finely pilose basally. Raceme cylindrical with rounded apex, compact, to 140 × 30 mm, with c.50 flowers. Bracts triangular, 1 mm long, membranous. Pedicels to 7 mm long, green. Flowers bell-shaped to spreading, 13 mm in diameter. Tepals equal to sub-equal, narrowly spathulate, to 9 × 2.5 mm, white with a violet to brownish-red mid-nerve. Filaments white, narrowly subulate, 5 × 1 mm; anthers oval, 1 mm long, with mauve walls. Ovary ovoid, 3 × 2 mm, white to pale green; style 2 mm long, white. Capsule emarginate, 12 × 9 mm, with persistent tepals. Seeds 6 mm long, with straight hairs up to 9 mm long.

Recorded from South Africa near the SW Botswana border (Kalahari Gemsbok Park, *van der Walt* 5792 (PRE)), but probably also found in Botswana. Also in Namibia and South Africa. Mostly in flat open areas of sandy soil, but in mountainous areas and cliff crevices in the N Cape; c.1000 m.

Conservation notes: If found within the Flora area probably Vulnerable D2, but not threatened globally.

8. **Eriospermum rautanenii** Schinz in Bull. Herb. Boiss. **6**: 522 (1898). Type: Namibia, Amboland, Olukonda, 23.xii.1896, *Rautanen* 227 (G lectotype, K, Z).

> *Eriospermum sphaerophyllum* Baker in F.T.A. **7**: 472 (1898). Type: Botswana, Ngamiland, near Kwebe, xii.1896, *E.Lugard & Lugard* 78 (K holotype).
> *Eriospermum currori* (Baker) Baker in F.T.A. **7**: 474 (1898). Type: Angola, no locality, *Curror* 26 (K).
> *Eriospermum brevipedunculatum* Poelln. in Feddes Repert. **52**: 125 (1943). Type: Namibia, Hereroland, Etiro, *Rautanen* 443 (Z).

Eriospermum rubromarginatum Poelln. in Feddes Repert **53**: 185 (1944). Type: Namibia, Ausberge, Farm Lichtenstein, 22.ii.1923, *Dinter* 4497 (B).

Plants with leaves appearing after the flowers, inflorescence up to 90 mm high. Tuber sub-globose to turbinate or multiple, to 30 × 45 mm; skin brown, internal flesh white; growing point apical. Old leaf sheaths 25–70 × 3 mm, light brown, finely fibrous or membranous. Leaf solitary, erect; contemporary leaf sheath exserted to 20 mm, to 2 mm wide, reddish-green, glabrous; lamina ovate to orbicular, to 60 × 40 mm, green to glaucous, coriaceous, glabrous with a distinct white marginal rim, sometimes crisped. Peduncular bract leaf-like, sheathing part to 30 × 4 mm, white glabrous, blade spreading to horizontal, ovate to orbicular-cordate, mucronate, coriaceous, to 62 × 85 mm, green to glaucous on both surfaces, glabrous. Peduncle to 15 × 2 mm, glaucous. Raceme compact, sub-corymbose, 35 × 45 mm, with 15–35 flowers. Bracts triangular, 1.5 mm long, closely appressed to underside of pedicels, green with transparent margin. Lowest pedicels 26 mm long, upper 10 mm, spreading to spreading-erect. Flowers spreading to recurved, to 10 mm in diameter. Tepals equal to sub-equal, white with a green mid-nerve and faint red streaking; outer narrowly elliptic, 5 × 2 mm, inner narrowly obovate, 5.5 × 2 mm. Filaments filiform, outer 3.25 mm long, inner slightly longer, 1 mm at widest, white; anthers pale yellow, less than 0.5 mm long. Ovary ovoid, 2.5 × 1.5 mm, pale green; style 2.5 mm long, white. Capsule to 9 mm long × 8 mm wide with uneven locule development; tepals persistent. Seeds 5 mm long, brown, hairs c.4 mm long.

Botswana. N: Ngamiland, c.7 km NW of Kwebe Hills, fl. 25.xii.1977, *P.A. Smith* 2147 (BOT, K). SW: Ghanzi Pan, 14 km E of Ghanzi, fl. 21.x.1969, *Brown* 6993 (K, SRGH); Okwa Valley, 7 km E of Gobololo, fl. 2.ii.1979, *Kreulen* 589 (BOT, PRE, SRGH).

Also in Angola and N and C Namibia. Pebbly quartz or clay soils; c.1100 m.

Conservation notes: A moderately widespread species; probably Lower Risk near threatened.

9. **Eriospermum flagelliforme** (Baker) J.C. Mannning in Bothalia **30**: 157 (2000). Type: Ethiopia, Gallabat, Gendua, 1865, *Schweinfurth* 26 (K lectotype, BM, G); lectotypified by Perry (1994).

Anthericum flagelliforme Baker in J. Bot. **1**: 140 (1872). Type: South Africa, Apies (Aapages) R., *Burke* s.n. (K holotype).

Schizobasis flagelliformis Baker in J. Linn. Soc., Bot. **15**: 261 (1877). Type as above.

Eriospermum abyssinicum Baker in J. Linn. Soc., Bot. **15**: 263 (1877); in F.T.A. **7**: 471 (1897). —van Wyk & Malan, Field Guide Wild Fl. Witwaters.: 152 (1988). —Whitehouse in F.T.E.A., Eriospermaceae: 4 (1996). Type: Ethiopia, Gallabat, Gendua, 1865, *Schweinfurth* 26 (K holotype, BM, G).

Eriospermum burchellii Baker in Fl. Cap. **6**: 372 (1896). Type: South Africa, Griqualand West, near Asbestos Mts, between Wittewater & Rietfontein, 15.ii.1812, *Burchell* 2008 (K lectotype).

Eriospermum luteorubrum Baker in Fl. Cap. **6**: 372 (1896). Type: South Africa, Transvaal, Barberton, summit of Saddleback Range, x.1889, *Galpin* 528 (K lectotype, BOL, PRE, SAM, Z).

Eriospermum fleckii Schinz in Bull. Herb. Boiss., ser.1 **4**, app.3: 37 (1896). Type: Namibia, Rehoboth, *Fleck* 887 (Z holotype).

Eriospermum elatum Baker in F.T.A. **7**: 471 (1898). Type: Zambia, Urungu, 1894, *Carson* 16 (K lectotype).

Eriospermum schinzii Engl. & Krause in Bot. Jahrb. **45**: 140 (1910), non Baker. Type: Namibia, Grootfontein, 30.xi.1908, *Dinter* 923 (SAM).

Plants with leaves appearing after the flowers; inflorescence 300–400 mm high. Tuber solitary, like a flattened globe, to 42 × 40 mm; skin dark brown, thick, rough, internal flesh white, corky; growing point apical. Old leaf sheaths to 60 × 10 mm, dark reddish-brown, fibrous, bristly on top. Leaf solitary, erect; contemporary leaf sheath terete, exserted to 130 mm, 1–2 mm in diameter, straight or lightly coiled at base, glabrous; lamina narrowly lanceolate, sometimes falcate, cuneate, acuminate, mucronate, up to 140 × 17 mm, glaucous green, glabrous, coriaceous, with prominent parallel veins; margin thickened or involute. Peduncular bract not observed. Peduncle to 180 × 1 mm, lower 15 mm dark red, remainder green, glabrous. Raceme very lax, 90–320 × 100 mm, with up to 50 flowers. Bracts triangular, 1–2 mm long, membranous with a brown mid-nerve. Lowest pedicels to 180 mm long, upper 10–20 mm, erect-spreading, arcuate. Flowers bell-shaped, becoming spreading to recurved in bright sun. Tepals sub-equal, lemon to bright yellow, with green mid-nerve overlaid with red streaking, outer strap-shaped to elliptic, 3–10 × 1.5 mm, inner spathulate to boat-shaped, 3–10 × 1–2 mm. Filaments sub-equal, filiform to narrowly lanceolate, 2.25–3 mm long, yellow; anthers small, ovoid. Ovary ovoid, 2–2,5 × 1.25 mm, pale green with some red stippling; style 1 mm long, yellow. Capsule turbinate, 8–9 mm long × 5–7 mm wide. Seeds c.6 × 2 mm; hairs dense 7 mm long, silvery-white when fresh.

Botswana. SW: Ghanzi Dist., Groote Laagte, *P.A. Smith* 3175 (BOT, PRE, SRGH). SE: Sebele, 5 km N of Gaborone, 1030 m, fl.& fr. 30.x.1977, *Hansen* 3260 (K, PRE, SRGH). **Zambia**. B: Kaoma Dist., c.80 km W of Kafue Hoek, fl. 7.xi.1959, *Drummond & Cookson* 6215 (LISC, SRGH). N: Kasama Dist., Chambeshi, Kalungu R., 4.i.1963, *Astle* 1888 (SRGH). W: Mwinilunga Dist., c.11 km NW of Kalene Mission, 1300 m, fl. 11.xi.1962, *Richards* 17172 (BR, LISC). C: Serenje Dist., Kundalila Falls, fl. 13.x.1963, *Robinson* 5719 (SRGH). S: Namwala Dist., Kafue Nat. Park, fl. 6.xii.1960, *Mitchell* 2/16 (SRGH). **Zimbabwe**. N: Gokwe Dist., Chimvuri vlei, c.3 km W of Gokwe, fl. 12.xi.1963, *Bingham* 931 (BR, LISC, LMU, SRGH). W: Matobo Dist., Besna Kobila Farm, fl. 28.xi.1962, *Miller* 8309 (SRGH). C: Harare Dist., Beatrice vleis, fl. 29.xii.1964, *Drewe* 4/65 (SRGH). E: Chimanimani Mts, fl. 16.xi.1967, *Mavi* 683 (LISC, SRGH). **Malawi**. N: Nyika Plateau, fl. 19.xi.1962, *Cottrell* 36 (SRGH). C: Dedza Dist., Dedza, 1700 m, fl. 18.i.1959, *Robson* 1246 (K, LISC, SRGH). S: Zomba Dist., Balaka, Toleza Farm, fl. 4.xii.1956, *Jackson* 2088 (BR, K, SRGH). **Mozambique**. Z: Mocuba Dist., 7 km from Mocuba–Mugeba turn-off to Maganja da Costa, 130 m, fl. 20.xi.1967, *Torre & Correia* 16141 (LISC). MS: Manica Dist., Chimoio, Vandúzi, fl. 5.i.1948, *Mendonça* 3623 (LISC).

Also in Senegal, Nigeria, Central African Republic, Sudan, Ethiopia, Kenya, Uganda, Tanzania, Namibia and South Africa. Varied habitats but principally grassland, also miombo and mopane woodland, usually on sandy or well-drained soils among rocks; 50–1800 m.

Conservation notes: Widespread species; not threatened.

The most common and widespread species of *Eriospermum*, and highly variable with many local variants.

10. **Eriospermum kiboense** K. Krause in Bot. Jahrb. **48**: 356 (1912). Type: Tanzania, Kilimanjaro, xii.1910, *Endlich* 711 (M holotype). FIGURE 13.2.3.

   Eriospermum erectum Suess. in Trans. Rhod. Sci. Assoc. **43**: 74 (1951). Type: Zimbabwe, Marondera, 8.xi.1941, *Dehn* 449 (M).

Plant with leaves appearing after the flowers; inflorescence up to 500 mm high. Tuber irregular in shape, to 50 × 35 mm; internal flesh white; growing point basal or lateral. Old leaf sheaths to 60 × 8 mm membranous, brown. Leaf solitary, erect; contemporary leaf sheath exserted for 50–60 mm, 3 mm in diameter, reddish-purple, glabrous; lamina broadly elliptic to lanceolate, to 140 × 90 mm. Peduncular bract up to 55 mm above ground, broadly ovate, to 25 × 18 mm, green, glabrous. Peduncle to 450 × 2–3 mm, glabrous. Raceme cylindrical, to 160 × 30 mm, with 20–50 flowers. Bracts 1 mm long, membranous. Pedicels to 10 mm long, spreading-erect. Flowers rotate, to 10 mm in diameter. Tepals c.5.5 × 2 mm narrowly oblong with obtuse to rounded apex, white to very pale green with darker green to brown mid-nerve. Filaments erect forming a ring round ovary, 2.5–3 × 1.2 mm, oblong, narrowing to a 0.2 mm neck supporting the anther. Ovary ovoid to globose, 1.75 × 1.75 mm, bright green; style cylindrical 1.5 mm long, white. Capsule to 9 × 9 mm, with persistent tepals. Seeds 2–3 mm long, with hairs up to 7 mm long.

Malawi. N: Nyika Plateau, 29.iii.1991, *Perry* 3810 (MAL, NBG); Mzimba Dist., Viphya Mts, Champira forest, fl., 26.xii.1975, *Pawek* 10543 (K, MAL, SRGH). **Zimbabwe**. C: Harare, 1460 m, fl. 16.xii.1926, *Eyles* 4575 (K, SRGH); Gweru, Gweru Teachers College, c.1400 m, fl. 29.xi.1966, *Biegel* 1484 (BR, K, LISC). E: Mutare Dist., Nyamashuta R., c.1000 m, fl.& fr. 29.xi.1952, *Chase* 4730 (K, MO, SRGH).

Also in Tanzania. In grassland and open woodland; 1000–1800 m.

Conservation notes: A moderately widespread species, although with a disjunct distribution; not threatened.

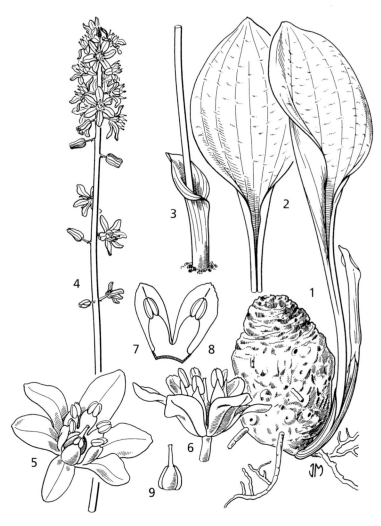

Fig. 13.2.**3**. ERIOSPERMUM KIBOENSE. 1, plant with tuber and leaf (× ²/₃); 2, upper surface (× ²/₃); 3, base of peduncle and peduncular bract (× ²/₃); 4, inflorescence (× ²/₃); 5, flower face view (× 3), 6, flower side view (× 3); 7, outer tepal and stamen (× 3); 8, inner tepal and stamen (× 3); 9, gynoecium (× 3). All from *Lavranos* 22848. Drawn by J. Manning. From Contrib. Bolus Herb. (1994).

Species for which material has not been traced

Eriospermum bussei Dammer in Bot. Jahrb. Syst. **38**: 64 (1905). Type: Malawi, Melingoti, fl. xii.1900, *Busse* 714 (presumed lost).

Flowers apparently produced before leaves as tuber and leaf are not included in the description. Nothing in the description relates it to any known species.

Eriospermum longipetiolatum Dammer in Bot. Jahrb. Syst. **38**: 64 (1905). Type: Malawi, Melingoti, fl.xii.1900, *Busse* 698 (presumed lost).

Tuber and flowers are missing from the description. Possibly the same as *E. bussei*, and both may be a form of *E. abyssinicum*.

Eriospermum peteri Poelln. in Feddes Repert. Spec. Nov. Regni Veg. **52**: 120 (1943). Type: Zimbabwe, Forest Wall between Bulawayo and Pasipas, fr. xii.1913, *Peter* s.n. (presumed lost).

Leaves and flowers appearing at the same time. Not possible to identify from the description.

Eriospermum pilosum Poelln. in Feddes Repert. Spec. Nov. Regni Veg. **52**: 121 (1943). Type: Botswana, Palapye, fl. viii.1896, *Passarge* 113 (presumed lost).

From the description this is probably a form of *E. porphyrovalve*.

Eriospermum seineri Engl. & K. Krause in Bot. Jahrb. Syst. **45**: 140 (1910). Type: Botswana, SW of Kuke pan, fl. i.1907, *Seiner* 12/332 (presumed lost).

The type specimen was found in SW Botswana in deep red sand, apparently flowered in January 1907. From the description it appears to be a form of the ubiquitous and variable *E. flagelliforme*, although von Poellnitz (1944) made it a variety of *E. bakerianum* Schinz. Not recorded from elsewhere.

Uncertain species and uncertain identifications

Eriospermum andongense Baker in Trans. Linn. Soc., Bot. **2**: 262 (1878). Type: Angola, Huìla, vi.1878, *Welwitsch* 3757 (BM lectotype, K).

Reported from Zimbabwe in the Sabonet checklist (Mapaura & Timberlake, Checklist Zimb. Fl. Pl.: 90, 2004), probably in error; no specimens seen. Otherwise only known from Angola.

Eriospermum cooperi Baker in J. Linn. Soc. **15**: 265 (1876). Type: Lesotho, 1861, *Cooper* 3307 (K holotype).

Reported from Zimbabwe in the Sabonet checklist (Mapaura & Timberlake, Checklist Zimb. Fl. Pl.: 90, 2004), but no specimens seen. Recorded from KwaZulu-Natal through to N South Africa. Perry (1994) says it may occur in Zimbabwe.

Eriospermum corymbosum Baker in J. Linn. Soc. **15**: 266 (1876). Type: South Africa, N Cape, Kimberley, Du Toit's Pan, xii.1872, *Tuck* in *Herb. MacOwan* 1969 (K lectotype, GRA).

Reported from Botswana in the Sabonet checklist (Setshogo, Prelim. Checklist Pl. Botswana: 122 (2005). No specimens seen, but known to occur close to the S Botswana border (Perry 1994).

Eriospermum flexuosum Baker in Trans. Linn. Soc. **2**: 261 (1878). Type: Angola, Huíla, between Lopollo and Humpata, vi.1878, *Welwitsch* 3760 (BM lectotype, K).

The type and a few other specimens contain slender inflorescences with a distinctly flexuose rachis. None of the specimens seen include leafing material, indicating that leaf appears separately from flowers. This could be a variant of *E. flagelliforme*, but without detail of leaves its status must remain in doubt. The name has also been used for other specimens in herbaria in both Zimbabwe and Zambia.

Zambia. N: Mbala Dist., rocky hill above Ndundu, 1500 m, fl. 16.ii.1957, *Richards* 8195 (BR, K, LISC). C: Mkushi, c.1500 m, fl. 7.i.1958, *Robinson* 2669 (K).

Eriospermum homblei De Wild. & Ledoux in De Wildman, Contrib. Fl. Katanga, Suppl.**3**: 96 (1930). Type: Congo, Katanga, Lualaba, by kraal, xii.1912, *Homblé* 960 (BR holotype).

Apparently an extra robust form of *E. mackenii* subsp. *mackenii* with up to 7 leaves.

Zambia. W: Mwinilunga Dist., source of Matonchi Dambo, fl. 16.xi.1937, *Milne-Redhead* 3257 (BR, K); Mwinilunga, c.5 km SE of Angolan border, 1300 m, fl. 7.xi.1962, *Richards* 16912 (K).

Eriospermum porphyrium Archibald in J.S. Afr. Bot. **26**: 115 (1960) Type: South Africa, Cape Prov., above Bushman's River Poort, *Archibald* 7199 (GRA holotype).

Reported from Botswana in the Sabonet checklist (Setshogo, Prelim. Checklist Pl. Botswana: 122 (2005), but no specimens seen. Known from South Africa (Eastern Cape, Northern Cape, Free State and former Transvaal).

DRACAENACEAE

by I. la Croix

Trees, shrubs (sometimes scandent), suffrutices or rhizomatous geophytes, from 10 cm to over 40 m tall. Trunks and aerial stems often with characteristic persistent leaf scars. Leaves alternate, distichous or spirally arranged, entire, ovate, lanceolate, oblanceolate, ligulate or ensiform, succulent or coriaceous, often with a sheathing base, up to 2 m long but usually much shorter, sometimes variegated. Inflorescence terminal, branched or unbranched. Flowers often in clusters, usually white but also yellowish or flushed with green or purple, scented. Pedicel articulated, with a part remaining on the inflorescence after flowers or fruit fall. Perianth with long or short tube and 6 equal segments. Stamens 6, of similar length to free perianth lobes, inserted at their base on the tube; filaments sometimes inflated and spindle-shaped; anthers versatile, latrorse. Ovary superior, 3-locular, each with a single ovule; septal nectaries present; style of equal length to stamens; stigma capitate or shallowly 3-lobed.

Two genera with most species in tropical and subtropical regions of the Old World but a few in Central America, Cuba, Hawaii, Cape Verde and the Canary Is. Widely introduced and sometimes naturalised elsewhere.

Dracaenaceae has been included under Agavaceae by some earlier authors, and is included under Asparagaceae in the recent on-line Kew World Monocot checklist. However, here it is kept separate.

The flowers of *Dracaena* and *Sansevieria* are virtually identical and some authors, notably Bos (in Kubitzki, Fam. Gen. Vasc. Pl. **3**: 238, 1998), consider that it is not possible to separate them and that *Sansevieria* should be included in *Dracaena*. However, there are differences in pollen morphology and in the fruit. The fruit of

Dracaena is a globose or ellipsoid berry with three large seeds, although often only one of these develops. In *Sansevieria*, the ovary is also trilocular with a single ovule in each, but as the fruit develops the ovary wall falls away and the seeds develop a fleshy covering, a sarcotesta, so that although the fruit resembles a berry, it is not a true berry. In view of this, until further studies, including DNA analysis, have been carried out, it seems more appropriate to keep the two genera separate.

Trees, shrubs or lianas; leaves flat, not fibrous and not greatly thickened; fruit a berry . . . . **1. Dracaena**
Stemless plants with a creeping rhizome; leaves flat, boat-shaped or cylindrical, succulent; fruit with thin pericarp falling away from berry-like seeds . . . .
. . . . **2. Sansevieria**

1. DRACAENA L.

Dracaena L., Syst. Nat., ed.12 **2**: 246; Mant. Pl.: 9, 63 (1767).
Pleomele Salisb., Prodr. Stirp. Chap. Allerton: 245 (1796). —N.E. Brown in Bull. Misc. Inform. Kew **1914**: 273 (1914).

Trees, shrubs or lianas with woody stems, rarely rhizomatous herbs; roots usually orange. Leaves ± sessile, usually spirally arranged, often in rosettes near ends of branches. Inflorescences terminal on branches, usually paniculate but sometimes racemose. Pedicels articulated in middle or near apex. Flowers in clusters of one to several, white or greenish-white, occasionally purple-tinged, usually fragrant and nocturnal; perianth with a long or short tube and 6 free lobes. Stamens 6, inserted on tube below junction with lobes. Ovary superior; stigma capitate. Fruit a globose or ellipsoid berry, usually orange or red when ripe, with 3 large seeds, although frequently only one develops. Although a few species of *Dracaena* are rhizomatous herbs, all occurring in the Flora area are trees, shrubs or lianas.

Genus of about 60 species, mainly in tropical Africa but also in Madagascar, Arabian peninsula, Socotra, SE Asia, Central America and the West Indies.
There are unconfirmed reports (Sabonet Zambia checklist, 2004) of *Dracaena aubryana* Brogn. in N & W Zambia, a West African species also recorded from E Uganda.

1. Trees or ± erect shrubs 2
 – Scandent or scrambling shrubs or lianas 7
2. Leaves less than 10 cm long; inflorescence unbranched . . . . **8.** *reflexa*
 – Leaves over 10 cm long; inflorescence branched 3
3. Leaves up to 30 cm long 4
 – Most leaves more than 30 cm long, usually much more 5
4. Leaves scattered along branches; perianth 40–50 mm long . . . . **1.** *mannii*
 – Leaves crowded at ends of branches; perianth 12–15 mm long . **5.** *afromontana*
5. Large unbranched or sparsely branched shrub 1–4 m tall (in Flora area); inflorescence to 60 cm long; perianth 12–25 mm long . . . . **2.** *fragrans*
 – Trees 6–18 m tall; inflorescence c.1 m long 6
6. Perianth 10–12 mm long; berries 12–17 mm diameter . . . . **3.** *steudneri*
 – Perianth 20–40 mm long; berries 8–10 mm diameter . . . . **7.** *aletriformis*
7. Leaves ± distichous; inflorescence branched, flowers borne singly . . **6.** *laxissima*
 – Leaves crowded at ends of branches; inflorescence unbranched; flowers in dense clusters . . . . **4.** *camerooniana*

1. **Dracaena mannii** Baker in J. Bot. **12**: 164 (1874); in F.T.A. **7**: 438 (1898). — Hepper in F.W.T.A. ed.2, **3**: 156 (1968). —Bos in F.S.A. **5**(3): 3 (1992). —Venter in Aloe **33**: 62 (1996). —White *et al.*, Evergr. For. Fl. Malawi: 99 (2001). —M. Coates Palgrave, Trees Sthn. Africa: 115 (2002). —Mwachala in F.T.E.A., Dracaenaceae: 7 (2007). Type: Nigeria, Old Calabar R., 1863, *Mann* 2329 (K lectotype, B, P, WAG). FIGURE 13.2.**4**.

Dracaena nitens Baker in Trans. Linn. Soc., Bot. **1**: 252 (1877). Types: Angola, Pungo Andongo, xii.1856, *Welwitsch* 3741 (BM syntype, K); same locality, *Welwitsch* 3742 (BM syntype, K) & Golungo Alto, ix.1857, *Welwitsch* 3743 (BM syntype, COI, K, P).

Dracaena reflexa Lam. var. *nitens* (Baker) Baker in F.T.A. **7**: 441 (1898). —White, For. Fl. N. Rhod.: 16 (1962). —K. Coates Palgrave, Trees Sthn. Africa: 87 (1977).

Dracaena usambarensis Engl. in Abh. Preuss. Akad. Wiss. 1894: 30 (1894); in Pflanzenw. Ost-Afr. **C**: 144 (1895). Type: Tanzania, Lushoto Dist., Mto wa Simbili beween Magila & Sigi R., *Volkens* 65 (B† syntype, K photo); Kilimanjaro, Schira, *Volkens* 1938 (B† syntype, BR, K photo).

Dracaena gazensis Rendle in J. Linn. Soc., Bot. **40**: 214 (1911). Type: Mozambique, Mt Maruma, 1905, *Swynnerton* 80b (SRGH holotype, K, BM).

Sparsely branched shrub or small tree to c.15 m high (trees can reach 50 m in West Africa). Leaves 10–30 × 0.8–2.2 cm, narrowly lanceolate to oblanceolate, acute, midrib slightly raised below, arranged along branches rather than in tufts. Inflorescence terminal on branches, pyramidal in outline, 12–38 cm long, 12–18 cm wide at base, with c.10 ± erect branches 6–9 cm long. Flowers very numerous, in fascicles of 2-4 on branches of inflorescence, cream or yellow-green with a broad white margin, very strongly scented. Perianth 40–50 mm long, including tube 18–30 mm long; tepals 20–30 × 0.5–1 mm, curling when flower opens. Stamens inserted near top of tube, slightly shorter than perianth; filaments 17–18 mm long, white, slightly flattened but narrowing where anther is joined; anthers yellow, 3 mm long. Ovary c.3 × 1 mm, showing as a small swelling at base of tube; style very slender, 38–50 mm long, ± equal or slightly longer than tepals; stigma small, capitate. Fruit fleshy, ± spherical, 7–20 mm diameter.

Zambia. N: Mbala Dist., Isoko Valley, Mwambeshi R., 900 m, fl. 5.ix.1960, *Richards* 13199 (K); Samfya Dist., Lake Bangweulu, near Samfya Mission, fl. 20.viii.1952, *White* 3095 (K, FHO). W: Zambezi Dist., Chitokoloki, *Martin* 902 (FHO), cited in White (1962). **Zimbabwe**. E: Chipinge Dist., Chirinda, 1150 m, fl. 10.x.1905, *Swynnerton* 80a (K). **Malawi**. N: Nkhata Bay Dist., Mzuzu–Nkhata Bay road, 550 m, fl. 6.ix.1975, *Pawek* 10087 (K, MAL, MO, SRGH, UC). C: Ntchisi Dist., Chipata Mt, 1350–1500 m, *Chapman* 1761 (FHO). S: Mulanje Dist., Lauderdale Crater, st. 7.iv.1958, *Chapman* 546 (K). **Mozambique**. N: Mueda Dist., between Mueda and N'gapa, fr. 26.xi.209, *Luke* 13919 (EA, K, LMA). Z: Lugela Dist., Namagoa, fl. 22.x.1948, *Faulkner* 81 (K). MS: Chibabava Dist., Madanda Forest, ix–x.1911, *Dawe* s.n. (K). GI: Homoine Dist., Inhamússua, fl. vii.1936, *Gomes e Sousa* 1771 (K, LISC). M: Marracuene Dist., Ricatla, fl. x–xi.1918, *Junod* 404 (LISC).

Widespread in tropical Africa from Senegal east to Zanzibar, extending south to Angola and South Africa; Guineo-Congolian linking species. Transitional woodland, lowland, mid-altitude and lower montane evergreen forest, riparian forest and evergreen thicket; 20–1600 m.

Conservation notes: Widespread species; not threatened.

2. **Dracaena fragrans** (L.) Ker Gawl. in Bot. Mag. **27**: t.1081 (1808). —Hepper in F.W.T.A., ed.2 **3**: 157 (1968). —Bos in Edinb. J. Bot. **49**: 311 (1992). —White *et al.*, Evergr. For. Fl. Malawi: 99 (2001). —Mwachala in F.T.E.A., Dracaenaceae: 8 (2007). Type: Illustration t.4/2 in J. Commelin, Horti Med. Amstelod. **2** (1701).

Aletris fragrans L., Sp. Pl., ed.2: 456 (1762).

Sansevieria fragrans (L.) Jacq., Fragm. Bot.: 5, t.2/6 (1800). —N.E. Brown in Bull. Misc. Inform. Kew **1915**: 259 (1915).

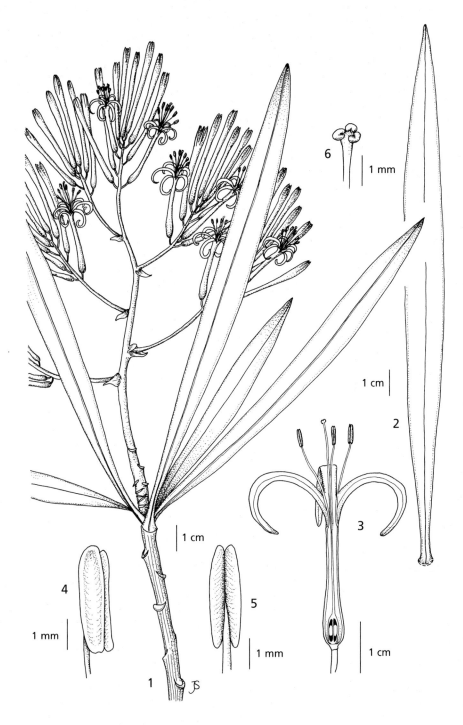

Fig. 13.2.4. DRACAENA MANNII. 1, habit; 2, leaf; 3, longitudinal section of flower; 4, anther, front view; 5, anther, back view; 6, stigma. All from *Dawe* 461. Drawn by Judi Stone.

Dracaena smithii Hook. f. in Bot. Mag.: t.6169 (1875); in F.T.A. **7**: 440 (1898). —Hepper in F.W.T.A., ed.2 **3**: 156 (1968). Type: Plant cultivated at Kew from West Africa (?Upper Guinea), i.1874 (K lectotype).

Dracaena deremensis Engl. in Bot. Jahrb. Syst. **32**: 95 (1903). Type: Tanzania, Lushoto Dist., Usambara, Handei, Nguelo, 9.vi.1899, *Scheffler* 67 (B† holotype).

Large shrub 1–4 m tall, sometimes forming large colonies. In the Flora area stems erect, almost always unbranched, but elsewhere can form a branched tree to 15 m tall. Leaves 20–80 × 4–6 cm, ovate, lanceolate or oblanceolate, acuminate, narrowed at base, with distinct midrib on underside. Inflorescence 25–55 cm long or more, usually with 6–10 branches 5–8 cm long near base. Flowers numerous, in well-separated, dense glomerules (usually more than 10 flowers in each) on branches and main axis, yellowish or white with pinkish lines, very strongly scented, turning black when pressed; bracts scarious, white, to 3 mm long; pedicel remnants 2–5 mm long; perianth 12–25 mm long, tube 5–12 mm long; tepals 7–13 mm long, 3 mm wide. Stamens inserted in throat, similar in length to perianth or slightly shorter, anthers c.2 mm long. Stigma 3-lobed, exserted for 1–3 mm. Fruit ± globose, sometimes lobed, 13–20 mm diameter, orange when ripe.

Zambia. N: Mporokoso Dist., Lumangwe Falls, fr. 14.xi.1957, *Fanshawe* 4020 (BR, K, NDO). **Zimbabwe**. E: Chipinge Dist., Chirinda Forest, 1150 m, fl. xii.1908, *Swynnerton* 6520 (BM, K). **Malawi**. N: Chitipa Dist., Misuku Hills, Mugesse Forest, 1570 m, fr. 7.vii.1973, *Pawek* 7065 (K, MAL, MO, UC). C: Ntchisi Forest Reserve, 1460 m, fl. 25.iii.1970, *Brummitt & Evans* 9388 (K, WAG). S: Thyolo and Mulanje Mts cited in White *et al.* (2001). **Mozambique**. MS: Gorongosa Dist., Mt Gorongosa, 1200 m, fr. 6.v.1964, *Torre & Paiva* 12297 (LISC). M: Maputo, Jardin Tunduru (Vasco da Gama), bud 16.ii.1972, *Balsinhas* 2388 (K, LMA).

Widespread in tropical Africa from Gambia to Ethiopia and south to Angola; a Guineo-Congolian linking species. Undergrowth in mid-altitude and montane evergreen forest, sometimes dominant (e.g. Chirinda Forest); 900–2000 m. Commonly cultivated in gardens, rooting readily from cuttings.

Conservation notes: Widespread species; not threatened.

A very variable species, as can be seen by the extensive synonymy given in Bos (1992).

3. **Dracaena steudneri** Engl., Pflanzen. Ost-Afr. **C**: 143 (1895). —Baker in F.T.A. **7**: 441 (1898). —White, For. Fl. N. Rhod.: 16 (1962). —K. Coates Palgrave, Trees Sthn. Africa: 87 (1977). —White *et al.*, Evergr. For. Fl. Malawi: 101 (2001). —M. Coates Palgrave, Trees Sthn. Africa: 116 (2002). —Mwachala in F.T.E.A., Dracaenaceae: 8 (2007). Type: Ethiopia, Gondar, Dschibba, 1862, *Steudner* 477 (B†, S, BR).

Dracaena papahu Engl., Pflanzen. Ost-Afr. **C**: 143 (1895). —Baker in F.T.A. **7**: 440 (1898). Type: Tanzania, Usambara Mts, Lutindi, x.1894, *Holst* 3260 (K isotype).

Tree 6–18 m high, with a few stout branches leaving trunk at an acute angle; trunk grey, to 30 cm or more in diameter. Leaves at ends of branches in whorls of c.30, older outer leaves drooping, inner ones erect, bases sheathing, 35–100 × 6.5–9.5 cm, lanceolate, acute, with thin cartilaginous margin, midrib visible below but less prominent than in *D. fragrans*. Inflorescence 1 m long or more, deltoid in outline, up to c.30 primary branches to 25 cm long, each with up to 6 secondary branches; bracts subtending branches papery, 25–30 mm long. Flowers in dense fascicles, dull white, pale green, yellow or yellow-green (creamy white when pressed), with a weak and rather unpleasant smell, said to be eaten by hyraxes; open flowers c.15 mm across; perianth 10–12 mm long, tube 4–5 mm long; tepals 5–8 × 0.8 mm; perianth persistent, twisted into a papery screw over the developing fruit. Stamens very slightly shorter than perianth; anthers c.2 mm long. Ovary 2–3 × 0.5 mm. Fruit globose, 12–17 mm diameter, red-brown to black when ripe.

Zambia. N: Chinsali Dist., Shiwa Ngandu (planted), st. 29.xi.1952, *White* 3777 (BM, FHO). **Zimbabwe**. C: Harare Dist., fl. vii.[n.d.], *Eyles* 7006 (K). E: Mutare Dist., Vumba Mts, forest slopes, 1400 m, fr. 1.xi.1946, *Wild* 1597 (K, SRGH). **Malawi**. N: Chitipa Dist., Misuku Hills, Mugesse Forest, 1850 m, fl. 1.i.1977, *Pawek* 12172 (K, MAL, MO, SRGH, UC). C: Ntchisi Dist., Ntchisi Mt, road from Forestry Station to mountain, 1500 m, fr. 2.v.1980, *Blackmore* 1372a (K, MAL). S: Blantyre Dist., Mt Soche, on NE slopes, 1400 m, fl. 26.viii.1982, *Chapman & Tawakali* 6390 (K, MAL). **Mozambique**. T: Angónia Dist., Mt Dómue, c.1600 m, fr. 9.iii.1964, *Torre & Paiva* 11094 (LISC). MS: Mossurize Dist?, Mt Maruma, 1060 m, fl. 11.ix.1906, *Swynnerton* 725 (BM, K).

Also in Sudan and Ethiopia south through E Congo to Mozambique. Evergreen montane and submontane forest; 700–2000 m. Often cultivated in gardens, rooting easily from cuttings.

Conservation notes: Widespread species; not threatened.

White *et al.* (2001) comment that this species and *D. fragrans* are very similar. However, the latter has shorter leaves, a different pattern of inflorescence branching and more brightly coloured fruits.

4. **Dracaena camerooniana** Baker in J. Bot. **12**: 166 (1874); in F.T.A. **7**: 442 (1898). —White, For. Fl. N. Rhod.: 16 (1962). —Hepper in F.W.T.A., ed.2 **3**: 157 (1968). —Mwachala in F.T.E.A., Dracaenaceae: 4 (2007). Type: Cameroon, Mt Cameroon, 1300 m, 1862, *Mann* 1204 (K holotype, A, P).

  Dracaena interrupta Baker in Trans. Linn. Soc., Bot. **1**: 252 (1877). Type: Angola, Pungo Andongo, by R. Taube, vi.1878, *Welwitsch* 3748 (K).

Erect or scrambling, much branched shrub 2–15 m tall. Leaves crowded at ends of branches, in tufts of 9–12, 10–22 × 2.5–5 cm, narrowly elliptic, elliptic or oblanceolate, acute or apiculate, gradually tapering to a sheathing base; midrib scarcely visible; first 5–6 nodes of new growth produce reflexed scarious scale leaves c.1 cm long. Inflorescence a lax raceme c.16 cm long. Flowers in dense clusters, set 1.5–6 cm apart, whitish-green or cream above, pink or reddish below, very fragrant; pedicels short, c.3 mm long; perianth 15–25 mm long, free segments shorter than tube, tube 10–16 mm long; tepals 7–10 mm long. Stamens of similar length to perianth or slightly shorter. Style slightly exserted. Fruit globose, 10 mm in diameter.

Zambia. N: Mpulungu Dist., Kamboli escarpment, 1500 m, fl. 13.vi.1961, *Richards* 15266 (K). W: Mwinilunga Dist., evergreen forest fringing Muzera R., 16 km W of Kakoma, bud 29.ix.1952, *White* 3405 (FHO, K).

Also in Sierra Leone, Ghana, Nigeria, Cameroon, Gabon, Angola. Forest undergrowth including swampy gallery forest, in high rainfall areas; 1200–1500 m.

Conservation notes: Common across N Zambia and elsewhere in Africa; not threatened.

5. **Dracaena afromontana** Mildbr., Wiss. Ergebn. Deutsch Zentr.-Afr. Exped., Bot. **1**: 62 (1910). —White *et al.*, Evergr. For. Fl. Malawi: 98 (2001). —Mwachala in F.T.E.A., Dracaenaceae: 6 (2007). Types: Rwanda, Rugege Forest, *Mildbraed* 1033; Ninagongo, *Mildbraed* 1360; Ruwenzori, *Mildbraed* 2525 (all B† syntypes). Neotype: t.5 in Mildbraed, Wiss. Ergebn. Deutsch Zentr.-Afr. Exped., Bot. **1**: 63 (1910), chosen by Bos (Fl. Ethiopia **6**: 76, 1997).

Sparsely branched shrub or shrubby tree 2–6 m high, sometimes scrambling to 8 m. Main trunk to 25 cm diameter. Leaves crowded towards ends of branches, in tufts of 12 or more, 15–30 × 1.5–3 cm, narrowly lanceolate or linear-lanceolate with faint midrib, broadly sheathing at base. Inflorescence pendulous, a lax panicle 20–60 cm long with curved branches; pedicels slender, remnants 4–9 mm long, some always more than 8 mm long. Flowers white or cream, up to 23 mm wide when open; perianth 12–15 mm long, tube c.1 mm long, free tepals 11–14 × 2–3 mm. Filaments swollen at base, c.9 mm long, inserted on tepals 3 mm above base, very slightly shorter than tepals; anthers 3 mm long. Ovary 4 × 2.5 mm, obovoid; style slightly exserted. Fruit globose, 15–18 mm diameter, orange.

Malawi. N: Chitipa Dist., Misuku Hills, Mugesse Forest, *Chapman* 1909 (cited in White *et al.* 2001); Rumphi Dist., Nyamkhowa Forest, 1800 m, st. ix.1902, *McClounie* 183 (K).

Also in Sudan and Ethiopia through E Congo and tropical East Africa; Afromontane endemic. In montane rainforest; 1700–2350 m.

Conservation notes: Within the Flora area restricted to forests in N Malawi; probably Vulnerable regionally.

6. **Dracaena laxissima** Engl. in Bot. Jahrb. Syst. **15**: 478 (1893). —Baker in F.T.A. **7**: 446 (1898). —White, For. Fl. N. Rhod.: 16 (1962). —Hepper in F.W.T.A., ed.2 **3**: 157 (1968). —White *et al.*, Evergr. For. Fl. Malawi: 99 (2001). —Mwachala in F.T.E.A., Dracaenaceae: 2 (2007). Type: Congo, Lund, Bashilange, near Mukenge, 18.ii.1882, *Pogge* 1462 (B holotype). FIGURE 13.2.**5**.

Slender, sparsely branched scandent shrub or liana to 10 m high, with long flexuous branches. Bark grey. Leaves in ± 2 rows, scattered along shoots, set 2–2.5 cm apart, 7–16 cm long × 2–5 cm wide, elliptic or lanceolate-elliptic, midrib prominent, narrowing to a pseudopetiole above broad sheathing base. Inflorescence pendulous, a slender panicle 15–30 cm long; bracts 2 mm long, subulate, inconspicuous. Flowers borne singly throughout inflorescence, greenish-white to pinkish, pendent on slender pedicels 15–20 mm long, articulated c.12 mm above base; perianth 15–22 mm long, segments twice as long as tube. Stamens of similar length to perianth; anthers white. Style slightly exserted. Fruit globose, orange, c.8 mm diameter, 1–3-locular.

Zambia. N: Isoka Dist., Mafinga Mts, 6 km W of Chisenga Resthouse, 2000 m, fl. 20.x.1952, *White* 3725 (FHO, K). E: Chama Dist., Nyika plateau, upper slopes of Kangampande Mt, 2100 m, fr. 7.v.1952, *White* 2755 (FHO, K). **Malawi**. N: Chitipa Dist., Jembya Forest Reserve, 18 km SSE of Chisenga, 1870 m, fr. 4.i.1989, *Thompson & Rawlins* 5976 (CM, K). C: Ntchisi Forest Reserve, 1640 m, fr. 25.iii.1970, *Brummitt & Evans* 9385 (K). S: Mt Mulanje, Lichenya plateau, Nessa path, 1800 m, fl. 23.xii.1988, *J.D. & E.G.Chapman* 9452 (K, MO). **Mozambique**. N: Ribáuè Dist., Serra de Ribáuè, Mepalué, 1350 m, fl. 25.i.1964, *Torre & Paiva* 10237 (LISC). Z: Gurué Dist., Serra do Gurué, 1200 m, fr. 24.ii.1966, *Torre & Correia* 14850 (LISC).

Also in São Tomé & Príncipe and Nigeria east to Zanzibar and Pemba and south to Mozambique. Montane and mid-altitude forest undergrowth; 850–2500 m.

Conservation notes: Widespread species, although restricted to moist forests; not threatened.

7. **Dracaena aletriformis** (Haw.) Bos in F.S.A. **5**(3): 3 (1992). —Venter in Aloe **33**: 62 (1996). —M. Coates Palgrave, Trees Sthn. Africa: 115 (2002). —Mwachala in F.T.E.A., Dracaenaceae: 6 (2007). Type: South Africa, E Cape, Uitenhage, 1840, *Drege* 4494a (K, G, MO, P), neotypified by Bos (Fl. Ethiopia **6**, 1992).

    Yucca aletriformis Haw. in Phil. Mag. J. **73**: 415 (1831).

    Dracaena hookeriana K. Koch in Wochenschr. Verein. Garten. Königl. Preuss. Staat. **4**: 394 (1861), illegitimate name. —Baker in J. Linn. Soc., Bot. **14**: 527 (1875); in Fl. Cap. **6**: 275 (1896). —K. Coates Palgrave, Trees Sthn. Africa: 86 (1977). Type: Illustration in Bot. Mag.: t.4279 (1847) from South Africa (designated by Bos 1992).

    Sanseveria paniculata Schinz in Durand & Schinz, Consp. Fl. Afr. **5**: 141 (1892), illegitimate name. Type: South Africa, E Cape, Port Alfred, *Schoenland* 290 (Z).

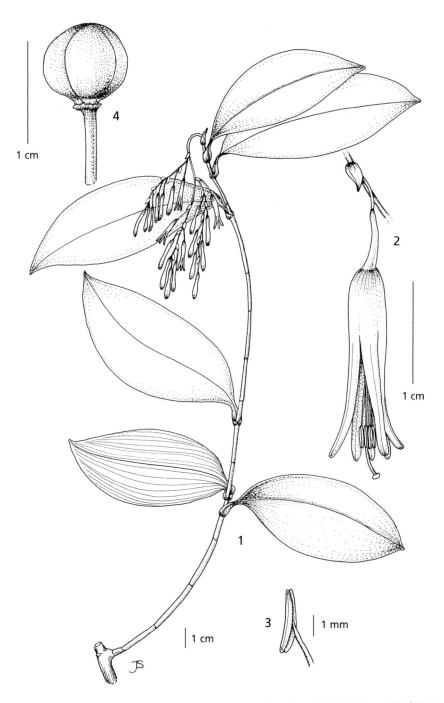

Fig. 13.2.**5**. DRACAENA LAXISSIMA. 1, habit; 2, flower; 3, anther; 4, fruit. 1–3 from *J.D. & E.G. Chapman* 9452, 4 from *Brummitt & Evans* 9385. Drawn by Judi Stone.

Small tree up to c.5 m tall, usually branched. Leaves borne towards ends of stems, 50–100 × 2.5–11cm, ± strap-shaped, acuminate, with narrow, white cartilaginous margins. Inflorescence erect, branched, to 1 m long. Flowers greenish-white, 1–4 in small clusters, 20–40 mm long; tepals c.1.5 times as long as tube; pedicel remnants 5–10 mm long. Fruit usually globose, 1-seeded, 8–10 mm wide, sometimes 2–3-lobed with 2–3 seeds and broader.

Mozambique. M: Namaacha Dist., Mt Ponduini, 740 m, st. 215.vii.1980, *Schafer* 7206 (K, LMU).

Also in South Africa, Swaziland, Kenya and Zanzibar. In coastal bush and forest; c.750 m.

Conservation notes: In the Flora area only known from S Mozambique; Vulnerable D2 in the FZ area, probably not threatened globally.

A variable species, both in size of leaf and size of flower.

Dracaena transvaalensis Baker is regarded as a separate species by Venter (1996). However, it is confined to South Africa and is not found in the Flora area.

8. **Dracaena reflexa** Lam., Encycl. **2**: 324 (1786). —Baker in J. Linn. Soc., Bot. **14**: 530 (1875). Type from Madagascar.

Small tree to 3 m tall, with few branches. Leaves in clusters at ends of stems, 7.5–9.5 × 0.8–1.5 cm, narrowly oblanceolate, acute to acuminate, grey-green when dry. Inflorescence unbranched, 12–13 cm long, flowers borne singly or in pairs. Flower colour not known. Pedicel remnants 2–3 mm long. Fruit 17–20 mm wide, bright orange.

Mozambique. N: Pemba Dist., Mt Ancuabe, 550 m, fr. 7.ii.1984, *de Koning & Groenendijk* 9517 (K, LMU); Nampula Dist., Namatuco, 350 m, fr. 30.i.1984, *Groenendijk, Maite & Dungo* 885 (K, LMU).

Also found in Madagascar. On basalt rocks with *Xerophyta, Myrothamnus* and *Aloe*, 350–550 m.

Conservation notes: Within the Flora area apparently only found in coastal N Mozambique; probably Vulnerable D2.

D. reflexa is an extremely variable species, widespread in Madagascar. Perrier (Fl. Madagascar **40**: 8, 1938) lists 14 varieties, which he says are only the principal ones, differing in many characters such as habit, size and shape of leaves, and form of inflorescence.

Most specimens of *D. reflexa* from Madagascar have longer and more slender pedicel remnants and smaller fruit than the specimens from Mozambique, but *Leeuwenberg* 14190 (Madagascar, Andohahela Reserve, Parcel Is., 500 m) agrees well with the Mozambique specimens. The flowers are cream, green at base, with perianth 14 mm long, tube 8–9 mm, and lobes 5–6 mm.

2. **SANSEVIERIA** Thunb.

Sansevieria Thunb., Prodr. Fl. Cap.: 65 (1794), conserved name —N.E. Brown in Bull. Misc. Inform. Kew **1915**: 185–261 (1915).

Evergreen perennial plants with creeping rhizomes. Stem woody, sometimes branched near base, or more commonly absent. Leaves distichous or in rosettes, usually succulent, cylindrical, with or without a channel, flat or boat-shaped, sometimes with light and dark markings, fibrous. Inflorescence racemose or paniculate. Flowers one to several in a cluster, often opening at night, usually fragrant and lasting for only one day or night. Pedicels articulated near apex or around the middle. Flowers white or greenish-white, sometimes purple-tinged (one Malagasy species, *S. sambiranensis* H. Perr., has deep pink to red flowers). Perianth with a tube and 6 lobes, often reflexed or rolled back. Stamens 6, attached to tube near junction with lobes; filaments slender, anthers versatile. Ovary trilocular, with one ovule in each loculus; style slender, stigma capitate. Fruit with a thin pericarp that falls away; seeds develop a fleshy coating (sarcotesta) so resembling a berry, but no trace of style remnants at apex.

About 50 species, most in tropical and subtropical Africa but extending into the Mascarene and Comoros Is., Arabian peninsula, India, Sri Lanka and Burma.

At one time several species were widely cultivated in the tropics for fibre obtained from the leaves, the quality of which apparently varied from species to species. Early botanical interest was primarily based on this use. Many species are now grown as ornamental plants.

Identification of herbarium specimens can be difficult and uncertain as material is often inadequate. A further complication is that many of the type specimens were prepared from cultivated plants, whose origin is often unknown.

Although a few species of *Sansevieria*, e.g. *S. arborescens* Cornu and *S. ehrenbergia* Schweinf., have woody stems and paniculate inflorescences, all species known from the Flora area are acaulescent with a simple racemose inflorescence, either spike-like or capitate.

A recent study done of Zimbabwean *Sansevieria* by Takawira-Nyenya (in Ghazanfar & Beentje, Taxon. Ecol. Afr. Pl.: 61, 2006), differs in some respects from the treatment below.

Key to species

Leaf descriptions apply to adult plants; species with cylindrical leaves usually have flat leaves in a rosette at the seedling stage.

1. Leaves cylindrical or elliptical in cross-section, channelled or not 2
 – Leaves flat or boat-shaped in cross-section 8
2. Flowers in a ± capitate head 3
 – Flowers in a spike-like raceme 4
3. Leaves 1–2 per shoot; inflorescence less than 15 cm tall . . . . **18.** *stuckyi*
 – Leaves 8–10 per shoot; inflorescence usually more than 15 cm tall . . . .
   . . . . **17.** *sinus-simiorum*
4. Leaves 1–2 per shoot . . . . **3.** *canaliculata*
 – Leaves more numerous, in 2 rows or in basal rosette 5
5. Leaves in a basal rosette . . . . **6.** *downsii*
 – Leaves in 2 rows 6
6. Perianth 18–20 mm long . . . . **14.** *pearsonii*
 – Perianth 30–45 mm long 7
7. Leaves 150–180 cm long, not channelled; inflorescence 60–90 cm tall . . . .
   . . . . **5.** *cylindrica*
 – Leaves up to 90 cm tall, channelled along their length; inflorescence 28–35 cm
   tall . . . . **2.** *burdettii*
8. Leaves narrowing abruptly to form a distinct petiole 9
 – Leaves narrowing gradually to a channelled base, usually short 10
9. Perianth 55–60 mm long, tube 40 mm; flowers in clusters of 3–6 . . . **4.** *concinna*
 – Perianth 30–50 mm long, tube 20–25 mm; flowers in clusters of 1–2(4) . . . .
   . . . . **19.** *subspicata*
10. Flowers in a ± capitate or dense ovoid head 11
 – Flowers in a spike-like raceme 15
11. Leaves ± channelled at base, but flat in cross-section in apical half 12
 – Leaves U-shaped in cross-section 13

12. Leaves 1–2(3) per shoot, ± erect .... **10.** *kirkii*
– Leaves 4–6 per shoot, in basal rosette .... **11.** *longiflora*
13. Leaves 4–6 per shoot, 6–10 cm wide; peduncle to 15 cm long; perianth tube 100–120 mm long .... **16.** *scimitariformis*
– Leaves 1–3 per shoot; peduncle less than 7 cm long 14
14. Leaves 60–90 cm long; peduncle to 7 cm long; perianth tube 45–95 mm long .... **7.** *hallii*
– Leaves 25–45 cm long; peduncle 3 cm long; perianth tube to 150 mm long ... . **8.** *humiflora*
15. Perianth at least 80 mm long 16
– Perianth up to 50 mm long 17
16. Pedicel remnants 3–4 mm long; leaves tapering to an acute apex .. **12.** *longistyla*
– Pedicel remnants 15–18 mm long; leaf apex narrowed abruptly to a sharp point .... **15.** *pedicellata*
17. Leaves to 2 cm wide, rough to touch, sometimes folded .... **1.** *aethiopica*
– Leaves at least 3 cm wide, usually more, not rough to touch 17
18. Leaves in basal rosette, 2–8 per shoot; perianth 30–50 mm long .. **9.** *hyacinthoides*
– Leaves 1–3 per shoot, ± erect; perianth 28–30 mm long .... **13.** *metallica*

1. **Sansevieria aethiopica** Thunb., Prodr. Fl. Cap.: 65 (1794); in Fl. Cap.: 329 (1823). —N.E. Brown in Bot. Mag. **139**: t.8487 (1913); in Bull. Misc. Inform. Kew **1915**: 230 (1915). —Sölch in Prodr. Fl. SW Afr., fam.148: 1 (1969). —Obermeyer in F.S.A. **5**(3): 7 (1992). —Newton in Eggli, Illust. Handb. Succ. Pl. **1**: 261 (2001). —Mbugua in F.T.E.A., Dracaenaceae: 17 (2007). Type: South Africa, Cape Prov., near Uitenhage, *Thunberg* s.n. (UPS).

   Sansevieria caespitosa Dinter in Repert. Spec. Nov. Regni Veg. **23**: 228 (1926). Type: Namibia, Maltahöhe, *Dinter* 3148 (B).
   Sansevieria scabrifolia Dinter in Repert. Spec. Nov. Regni Veg. **30**: 85 (1932). Type: Namibia, Hereroland, Otavifontein, fl. 24.i.1925, *Dinter* 5377 (B holotype, PRE).
   Sansevieria zeylanica sensu Baker in J. Linn. Soc., Bot. **14**: 548 (1875); in Bull. Misc. Inform. Kew **1887**(5): 8 (1887), non Willd.

Rhizomatous perennial, forming large colonies. Rhizome 8–14 mm in diameter. Leaves in clusters of 3–15, in rosette, erect to somewhat recurved, linear, partly folded, 25–40 × 1–2 cm, green mottled with bands of paler green, white cartilaginous margin with inner line of red; apex white, spiny, to 15 mm long; epidermis of leaves minutely papillate so surface feels rough. Inflorescence 40–95 cm long, with a dense, spike-like pseudo-raceme 22–56 × 4–6 cm; peduncle 25–40 cm long, with several loose, papery bracts 4–9 cm long, sheathing at base; pedicel 5–10 mm long, articulated near middle, remnants 3–8 mm long. Flowers white, cream or greenish-cream, sometimes purple-tinged, in clusters of 4–6, opening only at night, subtended by bract 9–10 mm long; perianth 33–50 mm long, tube 20–35 mm long; tepals 11–20 mm long. Stamens as long or slightly longer than perianth. Style exserted 3–4 mm. Fruit 1–3-lobed, a single lobe 6–8 mm wide, red when ripe.

Botswana. N: Okavango, Moremi Wildlife Res., island on Gobega lagoon, 940 m, fr. 5.iii.1972, *Biegel & Gibbs Russell* 3855 (K, LISC, SRGH); Makgadikgadi, 4 km S of Tsoi Camp on Rakops road, fl. 17.xi.1978, *P.A. Smith* 2514 (K, PRE, SRGH). SW: ?Ghanzi Dist., sand-covered ridge, 950 m, fr. v.1969, *R.E. Brown* 6081 (K, SRGH). SE: Central Dist., Ilalamabelel–Mosu area, near Sua Pan, fr. 15.i.1974, *Ngoni* 333 (K, SRGH). **Zambia**. B: Senanga Dist., near Nangweshi, S of Senanga, st. 16.x.1953, *Codd* 8022 (K, PRE) [no flowers present but specimen appears to be this species]. **Zimbabwe**. W: Matobo Dist., Besna Kobila, 1450 m, fl. xi.1954, *Miller* 2524 (K, SRGH). C: Kwekwe Dist., Sebakwe Botanic Reserve, fl. 17.i.1962, *Wild* 5616 (K, LISC, SRGH). S: Beitbridge Dist., c.16 km W of Beitbridge, fl. 11.i.1961, *Leach* 10692 (K,

SRGH). **Mozambique**. Z: Mocuba Dist., c.32 km from Mocuba to Mugeba, 200 m, 5.iii.1966, *Torre & Correia* 15037 (LISC).

Also in South Africa and Namibia. In sand or sandy pockets on rock; 200–1450 m.

Conservation notes: Widespread species; not threatened.

The species has been cultivated in Europe since about 1700.

2. **Sansevieria burdettii** Chahin. in Brit. Cact. Succ. J. **18**: 132 (2000). —Newton in Eggli, Illust. Handb. Succ. Pl. **1**: 262 (2001). Type: Malawi, Chikwawa Dist., Kapuchira Falls, *Burdett* in *Chahinian* 318 (K holotype [not found], MO), not *Chahinian* 316 in error.

Stemless plant with an underground creeping rhizome c.4 cm in diameter. Leaves 3–6, to 90 cm long, 2.5 cm wide and 3 cm thick, basal part with concave channel as wide as leaf, upper part cylindrical in cross-section, waxy, dark green, no markings or only very faint cross-banding on young leaves, several longitudinal lines running from base almost to apex; leaf apex withered, spiny, with a narrow dark brown line at junction with rest of leaf. Inflorescence a spike-like pseudo-raceme, many flowered, much shorter than leaves. Peduncle 16–20 cm long, 6–10 mm wide at base; raceme 12–15 cm long. Flowers in clusters of 4–6, usually 5, white, tinged pinkish brown, fragrant, open in evening and at night; pedicels 4 mm long, jointed at base of tube; perianth tube 20–25 mm long, 1 mm wide, cylindrical but slightly inflated to 1.5 mm at base, lobes 18–20 × 2 mm, convolute. Filaments 20 mm long; stamens 3 mm long, exserted. Style and stigma 38–40 mm long, up to 5 mm longer than stamens.

Malawi. S: Chikwawa Dist., Kapichira Falls, river bank, fl. in cult. 22.xi.1974, *Brummitt* 10005 (K).

Not known elsewhere. In rocky, riverine woodland; c.100 m.

Conservation notes: Known only from the type locality in S Malawi; Data Deficient, but probably Vulnerable D2.

This species resembles *S. cylindrica* Bojer, but differs in that the channelled leaves lack cross-banding and the inflorescences are shorter.

3. **Sansevieria canaliculata** Carrière in Rev. Hort. **1861**: 449 (1861). —N.E. Brown in Bull. Misc. Inform. Kew **1915**: 224 (1915). —Newton in Eggli, Illust. Handb. Succ. Pl. **1**: 263 (2001). Type: Unknown locality, cultivated specimens in Paris Bot. Garden, *Bojer* s.n. (P).

   Sansevieria sulcata Baker in J. Linn. Soc., Bot. **14**: 549 (1875) [as synonym under *S. cylindrica* Bojer]; in F.T.A. **7**: 335 (1898). Type: Unknown origin, *Bojer* in Hort. Maurit. (K).
   Sansevieria schimperi Baker in F.T.A. **7**: 335 (1898). Type: Cultivated plant at Kew, originally from Somalia, 20.vi.1892, *Stace* s.n. (K holotype).
   Sansevieria pfenningii Mbugua in F.T.E.A., Dracaenaceae: 33 (2009). Type: Tanzania, Lindi Dist., Lake Litamba, 28.i.1935, *Schleiben* 5917 (MO holotype, K).

Stemless perennial herb with creeping rhizome c.13 mm thick. Leaves 1–2 per shoot, ± terete, rather slender, 35–100 cm long, 0.5–2 cm wide, dull green, with 5–6 longitudinal grooves, tapering to a sharp 3 mm long point. Flowering stem 12.5–13.5 cm long; peduncle 5 cm long with 2 bracts c.15 mm long; inflorescence a rather lax, spike-like pseudo-raceme. Flowers white, tinged green, in clusters of 3; pedicel 1–1.6 mm; perianth 30–42 mm long, tube 18–25 mm long, lobes 12–17 mm long.

Mozambique. N: Niassa/Cabo Delgado Prov., Mt Mkota, st. xi.1907, *Stocks* 142 (K); Palma Dist., Lake Nhica, c.40 km W of Palma, 10°41'45"S 40°12'06"E, 30 m, old fl. 15.xi.2009, *Goyder, Alves & Massingue* 6071 (K, LMA).

Also in Madagascar and the Comores. Sandstone outcrops; 50–100 m.

Conservation notes: Disjunct distribution; Data Deficient.

The short inflorescence of this species is characteristic.

4. **Sansevieria concinna** N.E. Br. in Bull. Misc. Inform. Kew **1915**: 233 (1915). —
Newton in Eggli, Illust. Handb. Succ. Pl. **1**: 263 (2001). Type: Mozambique, near
Beira, fl. 2.vi.1911, *Dawe* 1 (K holotype).

   Sansevieria subspicata Baker var. *concinna* (N.E. Br.) Mbugua in F.T.E.A., Dracaenaceae: 17
   (2007).

Stemless perennial herb with creeping rootstock c.1 cm thick. Leaves 2–5 per shoot,
somewhat recurved, 15–35 cm long including petiole, 1–5.5 cm wide, narrowly lanceolate to
lanceolate, green with transverse paler bands (not visible in dried specimens); margins green,
sharply pointed at apex, narrowing at base to a channelled petiole 3–8 cm long. Flowering stem
15–37 cm tall; peduncle 10–20 cm long, tinged and spotted with purple, with 3–4 bracts 20–35
mm long; raceme dense. Flowers large, creamy white, in clusters of 3–6; pedicel jointed just
below flower, bases 3–7 mm long; perianth 55–60 mm long, tube 40 mm long, lobes 15–20 mm
long. Stamens of similar length to perianth. Style exserted. Fruit 9–15 mm wide.

   Zimbabwe. S: Chiredzi (Ndanga) Dist., near Save R., 14.vi.1950, *Chase* 2429 (BM,
SRGH). **Mozambique**. MS: Dondo Dist., Macuti, fl. 23.iii.1960, *Wild & Leach* 5234 (K,
SRGH). GI: Chibuto, between Maniquenique and Xai-Xai road, fr. 6.vi.1957, *Barbosa
& Lemos* 7606 (LISC).
   Also in South Africa (KwaZulu-Natal) and coastal S Tanzania. Evergreen coastal
scrub; 5–400 m.
   Conservation notes: Restricted to dune scrub; probably Near Threatened.

5. **Sansevieria cylindrica** Hook. in Bot. Mag. **85**: t.5093 (1859). —Baker in J. Linn.
Soc., Bot. **14**: 549 (1875); in Bull. Misc. Inform. Kew **1887**: 9 (1887); in F.T.A. **7**:
335 (1898). —N.E. Brown in Bull. Misc. Inform. Kew **1915**: 217 (1915). Type:
t.5093 in Bot. Mag. **85**, drawn from living plant from Angola.

   Sansevieria angolensis Carrière in Rev. Hort. **1861**: 447 (1861). Type: Angola, Luanda,
   iii.1854, *Welwitsch* 3749 (LISU holotype).
   Sansevieria livingstoniae Rendle in J. Bot. **70**: 89 (1932). Type: Zambia, Victoria Falls,
   Livingstone Is., 6.viii.1929, *Rendle* 374 (BM holotype).

Stemless perennial herb with a stout, creeping rootstock to 1 cm thick. Leaves 3–5 per shoot,
in 2 rows, 150–180 cm long, c.3 cm wide in middle, stiffly erect, completely round in cross-
section, bases ovate and overlapping, tapering to a sharp apex, with c.5 longitudinal grooves.
Flowering stem 60–90 cm tall; peduncle 30–45 cm long; inflorescence a dense, cylindrical spike-
like pseudo-raceme up to 60 cm long; bracts 4–8 mm long, scarious. Flowers creamy or pinkish
white, in clusters of 4–6; pedicel 5–6 mm long, articulated just below middle so remnant is
shorter than deciduous part; perianth 33–42 mm long, tube 16–27 mm long, slender but
slightly enlarged at base, lobes 12–18 mm long, c.1 mm wide. Stamens of equal length to
perianth. Style shortly exserted.

   Zambia. S: Livingstone Dist., Zambezi R., Livingstone Is., fl. 6.viii.1929, *Rendle* 374
(BM). **Zimbabwe**. W: Matobo Dist., Mangwe R., st. 8.ix.1871, *Baines* s.n. (K). C:
Harare, Bemerton rectory, garden of F. Warre (presumably cultivated), fl. ii.1914,
Brown s.n. (K).
   Also in Angola. Habitat unknown; 1000–1400 m.
   Conservation notes: Disjunct distribution; probably Near Threatened.
   This species has a confused nomenclatural history (see Brown 1915).
   Var. *patula* N.E. Br. (in Bull. Misc. Inform. Kew **1915**: 218) is described from
cultivated specimens but was originally from Angola. It is said to differ in having
recurved, somewhat spreading leaves and slightly smaller flowers, but is probably no
more than a cultivar and is not recognised here.

6. **Sansevieria downsii** Chahin. in Brit. Cact. Succ. J. **18**: 133 (2000). —Newton in
  Eggli, Illust. Handb. Succ. Pl. **1**: 264 (2001). Type: Malawi, Njakwa Gorge,
  16.x.1975, *Downs* 1/75 (K holotype, MO). FIGURE 13.2.**6**.

  Sansevieria aff. *gracilis* N.E. Br. sensu Thiede in Sansevieria J. **2**: 29, fig.2 (1993).

Stemless perennial with creeping stolons covered in leaf sheaths, several ovate, bract-like
leaves, 4.8–7 × 2 cm at the base of each shoot. Leaves 6–14, in rosettes, recurved, spreading,
14–45 cm long with ovate base c.2.5 × 2.5 cm, suddenly narrowing to c.3 cm wide, cylindrical
but channelled for $^1/_4$ to $^4/_5$ of its length on inner surface, mid-green with no or very faint grey-
green cross-banding, longitudinally grooved; leaf apex spiny, withered, a brown line at its base.
Inflorescence a spike-like pseudo-raceme, up to 160 cm long, much longer than the leaves;
rachis 40–70 cm long, about twice as long as peduncle; peduncle with one acuminate bract c.2
cm long; pedicel 2–5 mm long, articulated below middle. Flowers cream, pearly grey or pale
purplish-green, in clusters of 3–10, usually set 1–2 cm apart, opening in the evening; perianth
20–24 mm long, tube c.10 mm long, lobes c.12 mm long. Anther filaments c.16 mm long,
stamens 3 mm long, exserted. Style and stigma c.23 mm long. Fruits 8 mm diameter.

Zambia. N: Kaputa Dist., Kaputa, fr. 17.x.1949, *Bullock* 1309 (K). **Malawi**. N:
Rumphi Dist., S Rukuru R., Njakwa Gorge, fl. 16.x.1975, *Downs* 1/75 (K, MO);
Mzimba Dist., *Brachystegia* woodland near Lunyangwa R., fr. 26.xi.1976, *Pawek* 11952
(K, MO).

Not known elsewhere. On termite mounds in dry woodland and *Acacia-Euphorbia*
bush; 1000–1500 m.

Conservation notes: Endemic to the NE part of Flora area; probably Vulnerable D2.

Some specimens in Zambian herbaria (and one at Kew – Zambia N: Mbala Dist.,
Mulungu R. swamp, 6.x.1956, *Richards* 6389 (K spirit)) are labelled as *Sansevieria
gracilis* N.E. Br. and are probably referable to *S. downsii*. Although *S. gracilis* is
recorded for F.T.E.A., it is not known from the Flora area.

7. **Sansevieria hallii** Chahin. in Sansevieria J. **5**: 7 (1996). —Newton in Eggli, Illust.
  Handb. Succ. Pl. **1**: 266 (2001). Type: Zimbabwe, Birchenough Bridge, 1967,
  from cultivated plant, *Chahinian* 634 from *Hall* 67/799 (MO holotype, UPS).

Stemless perennial with creeping rhizome 18–30 mm thick. Leaves 1–3 per shoot, 60–90 cm
long, 5 cm wide in middle, erect, recurved, boat-shaped in cross-section, folded when pressed,
margin red-brown, dark grey-green with inconspicuous transverse bands when young but with
no visible markings when mature. Flowering stem short, 12–18 cm long; peduncle to 7 cm long
with c.5 broadly ovate, acute bracts 14–33 × 14–16 mm. Flowers in dense, capitate raceme to 16
cm wide, tube purple, lobes cream or white; bracts 18–30 × 3–6 mm, perianth to 120 mm long,
tube 45–95 mm long, lobes recurved, 20–25 mm long. Filaments 27–32 mm long; anthers 3–4
mm, slightly exserted. Style 70–114 mm long, exserted for c.15 mm. Fruit c.10 mm diameter,
globose, wrinkled.

Zimbabwe. E: Buhera Dist., Birchenough Bridge, c.300 m, fl. 1967, *Hall* 67/799 as
Chahinian 634 (MO, UPS); Chimanimani Dist., near Umvumvumvu R., on
Mutare–Birchenough Bridge road, fl. 21.i.1960, *Leach* 10714 (K, SRGH). S: Zaka
Dist., 360 m, fl. 22.xii.1951, *Wild* 3711 (K, PRE, SRGH). **Mozambique**. T: Moatize
Dist., Lupata, st. 20.iv.1860, *Kirk* 281 (K).

Also in South Africa (N Limpopo). Rocky outcrops; 300–360 m.

Conservation notes: Near endemic to the eastern interior of the Flora area, but
widespread; probably Lower Risk near threatened.

This species resembles the East African *Sansevieria fischeri* (Baker) Marais, but is
much smaller with broader leaves narrowing towards the base, a flattened channel
almost as wide as the leaf, chestnut margins and whitish membranes, obtuse or round
at the tip, and with longer flowers.

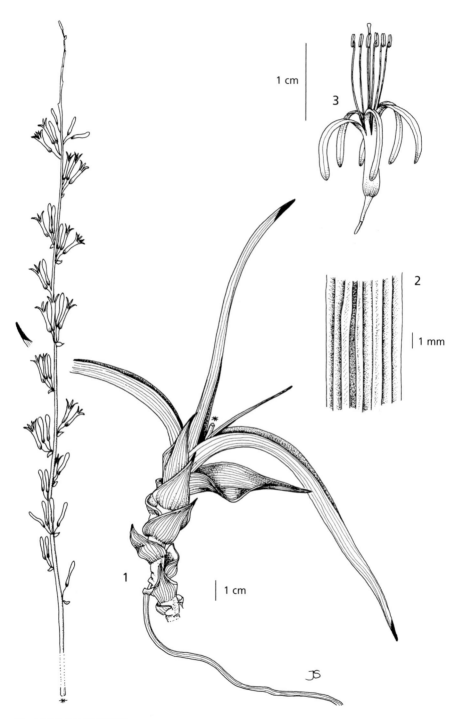

Fig. 13.2.**6**. SANSEVIERIA DOWNSII. 1, habit; 2, part of leaf showing ridges ; 3, flower. All from *Pawek* 10340. Drawn by Judi Stone.

8. **Sansevieria humiflora** D.J. Richards in Sansevieria **10**: 3 (2004). Type: Zimbabwe, near Mozambique border by Mt Selinda, from cultivated plant, 23.xi.1986, *Richards* R889 (SRGH holotype).

Robust, stemless plant forming tangled clumps. Rhizome stout, 2–3 cm thick. Leaves 1–3 per shoot, strap-shaped, recurved, rough textured and very rigid, 25–45 cm long, up to 8 cm wide and 10–15 mm thick at base, dark green on upper surface with numerous longitudinal grooves, deeper and more concentrated towards centre, under surface dark green, sometimes faintly banded or with indistinct lighter patches, with several shallow, longitudinal grooves; margins slightly wavy, with narrow brown line and thin fibrous stripe on extreme edge, tip blunt. Scape subterranean, 3 cm long, 1.5 cm wide, with basal bracts white or purplish if exposed to light and with numerous whitish streaks. Inflorescence a dense, capitate raceme, initially purplish in colour, emerging at ground level or slightly below, often hidden by leaves and humus, usually distorted and compressed by leaves, rarely free. Flowers purplish, in clusters of 4; bracts 12.5–15 × 10 mm, semi-transparent with purplish tips and many nerves; pedicel 3–4 mm long; perianth tube to 15 cm long, base slightly swollen, 4 mm wide, tapering to 2–3 mm in centre and slightly enlarged again towards top. Stamens and style exserted for at least 4 cm. Fruit globular, orange-red when mature, very wrinkled.

Zimbabwe. E: Chipinge Dist., near Mozambique border, 100 km NE of Chiredzi, 32°28'E 20°35.3'S, 23.xi.1986, *Richards* R889 (SRGH).

Known only from SE Zimbabwe. Dry bush, associated with *Aloe* and *Euphorbia* species; c.300 m.

Conservation notes: Known only from the type; Data Deficient, possibly Endangered.

Sansevieria humiflora differs from *S. hallii* in having a more compact growth habit and much shorter, thinner and flatter leaves lacking a central channel, and is a more uniform glossy green colour.

9. **Sansevieria hyacinthoides** (L.) Druce in Bot. Exch. Club Soc. Brit. Is. Rep. 1913, **3**: 423 (1914). —Obermeyer in F.S.A. **5**(3): 5 (1992). —Newton in Eggli, Illust. Handb. Succ. Pl. **1**: 266 (2001). —Mbugua in F.T.E.A., Dracaenaceae: 21 (2007). Type: Illustration t.33 in Commelin, Praeludia Bot. **84** (1703).

   Aloe hyacinthoides L., Sp. Pl: 321 (1753).
   Sansevieria thyrsiflora (Petagna) Thunb., Prodr. Pl. Cap.: 65 (1794). —N.E. Brown in Bull. Misc. Inform. Kew **1915**: 249 (1915). Type: Cultivated plant of unknown origin, possibly from Guinea.
   Sansevieria guineensis (L.) Willd., Sp. Pl., ed.4 **2**: 159 (1799). —Baker in F.T.A. **7**: 333 (1898).
   Sansevieria grandis Hook. f. in Bot. Mag. **129**: t.7877 (1903). Type: Illustration of cultivated specimen originally sent to Kew from Africa via Cuba, 1896.
   Sansevieria grandis var. *zuluensis* N.E. Br. in Bull. Misc. Inform. Kew **1915**: 252 (1915). Type: South Africa, KwaZulu-Natal, Somkele, cult. 24.i.1912, *Wylie* in NH 12010 (NH holotype).

Stemless perennial with creeping rhizome c.12 mm thick. Leaves in rosette, 2–8 to a node, 15–70 × 3–12 cm, lanceolate or broadly linear, erect, smooth, ± flat but narrowed and folded at base for c.5 cm, dull green with numerous pale transverse bands, margins whitish, with red-brown line. Inflorescence a spike-like pseudo-raceme, 45–82 cm tall, including a peduncle 17–53 cm long, 3–6 scarious bracts up to 5 × 2 cm; raceme c.7 cm wide. Flowers greenish-white, white or pale yellow tinged with brown or mauve, in irregular clusters of 2–6, opening in evening, scented; pedicels 4–6 mm long, articulated in middle or upper half (remnants c.4 mm long); perianth 30–50 mm long, tube 20–30 mm, lobes 14–20 mm. Stamens exserted. Style well exserted. Fruit a berry 6–10 mm wide, bright orange when ripe.

Zambia. B: Mongu Dist., Lukushi, 1050 m, st. vi.1933, *Trapnell* 1234 (K).

Zimbabwe. N: Makonde Dist., Mangula, Windale Farm, fr. 24.iii.1969, *Pope* 2 (K, LISC, SRGH). E: Mutare Dist., hillside near Quagga's Hoek, 1000 m, fl. 30.i.1955, *Chase* 5495 (BM, SRGH). S: Gutu/Buhera Dist., Ruti dam site, *Combretum*-mopane woodland, fl. viii.1974, *Ellert* in GHS 26672 (K, SRGH). **Malawi**. C: Nkhotakota Dist., N bank of Bua R. by bridge 19 km N of Nkhotakota, 475 m, fr. 16.vi.1970, *Brummitt* 11456 (K). S: Chikwawa Dist., Lengwe Game Reserve, 100 m, fr. 9.iii.1970, *Brummitt & Hall-Martin* 8991 (K). **Mozambique**. N: Nampula, Serra da Mesa, c.500 m, fr. 25.iv.1964, *Torre & Paiva* 12159 (LISC). Z: Lugela Dist., Namagoa, fr. 2.ii.1949, *Faulkner* 364 (K). MS: Marromeu Dist., Chiramba (Shiramba) & Chigogo (Shigogo), fr. i.1860, *Kirk* 178 (K). GI: Guijá Dist., Caniçado, fr. 3.vi.1959, *Barbosa & Lemos* 8587 (LISC). M: Magude Dist., Ungubane near Magude, fl. 5.xii.1980, *Jansen, Nuvunga & Petrini* 7665 (K).

Widespread in E tropical Africa and N & E South Africa. Sandy or stony soils, usually under shade of trees and bushes; 100–1050 m.

Conservation notes: Widespread species; not threatened.

Reportedly introduced into Dutch gardens before 1701. Cultivated in gardens in Zimbabwe (Biegel, Checklist. Orn. Pl. Rhod. Gdns.: 96, 1977).

Many specimens of *S. hyacinthoides* in regional herbaria and in checklists from the Flora area have been labelled as *Sansevieria conspicua* N.E. Br. It is certainly a variable species. Although *S. conspicua* occurs in the F.T.E.A. area, it is not clear whether it extends to the Flora Zambesiaca area or whether it is perhaps conspecific with *S. hyacinthoides* (see Takawira-Nyenya in Ghazanfur & Beentje, Taxon. Ecol. Afr. Pl.: 62, 2006) which has apparently shorter flowers.

10. **Sansevieria kirkii** Baker in Bull. Misc. Inform. Kew **1887**: 3 (1887); in F.T.A. **7**: 334 (1898); in Bot. Mag. **120**: t.7357 (1894). —N.E. Brown in Bull. Misc. Inform. Kew **1915**: 254 (1915). —Mbugua in F.T.E.A., Dracaenaceae: 30 (2007). Type: Tanzania, coast opposite Zanzibar, 1881 (fl. at Kew vi.1893), *Kirk* s.n. (K lectotype).

Stemless herb with stout rhizome. Leaves 1–2(3), erect or slightly spreading, 23–90 × 3–10 cm, narrowing to 1 cm at base, oblanceolate, very rigid, flat in upper half, channelled in lower half, 7–8 mm thick at halfway point, obscurely mottled with white, with a narrow red-brown marginal line; apex blunt, often disintegrates with red edge becoming detached. Peduncle 18–52 cm long, shorter than the leaves, with 5–6 large ovate bracts 6–10 cm long, 3–4 cm wide; rachis 2–7 cm long; raceme dense, capitate, 15–20 cm wide, subtended by several green, ovate, obtuse bracts, 3–3.5 × 1.6–2.2 cm. Flowers greenish-white with an unpleasant smell, in clusters of 6; pedicels 8–12 mm long, articulated just below the flower; floral bracts c.25 mm long; perianth 90–160 mm long, tube 80–125 mm long, lobes curling back, 20–35 mm long, 2–3 mm wide, oblanceolate. Stamens ± equal to tepals. Style exserted 20–50 mm. Fruit orange, 12 mm diameter.

Zambia. N: Mbala Dist., Inyendwe Valley, Lufuba R., 780 m, fl. 8.xii.1959, *Richards* 11898 (K). W: Ndola, termitaria in miombo, 1250 m, fl. 18.i.1957, *Fanshawe* 2943 (K, NDO). C: Lusaka Dist., Protea Hill Farm, 13 km SE of Lusaka, cult., fl. 25.xii.1994, *Bingham* 10223 (K). **Zimbabwe**. Recorded from N, W, C in Takawira-Nyenya (2006). **Malawi**. S: Shire Highlands, ii.1881, *Buchanan* 105 (K). **Mozambique**. T: Marávia Dist., Chicoa, Serra de Songa, c.900 m, fl. 31.xii.1965, *Torre & Correia* 13976 (LISC).

Also in Tanzania. Dry, rocky areas in woodland, on termite mounds and under bushes; 750–1500 m.

Conservation notes: Fairly widespread species; not threatened.

Introduced to cultivation by Sir John Kirk in 1881, who commented that it "... yields a most excellent fibre". It first flowered at Kew in 1893.

Var. *pulchra* N.E. Br. (Bull. Misc. Inform. Kew **1915**: 256, 1915) from Zanzibar is said to have more strongly variegated leaves, and is possibly just a cultivar. Brown (1915) says that the plant known as var. *pulchra* was received at Kew from the Paris Botanic Garden as *S. longiflora* Gérôme & Labroy, an illegitimate name.

11. **Sansevieria longiflora** Sims in Bot. Mag. **53**: t.2634 (1826). —Baker in J. Linn. Soc., Bot. **14**: 548 (1875); in Bull. Misc. Inform. Kew **1887**: 7 (1887); in F.T.A. **7**: 334 (1898). —N.E. Brown in Bull. Misc. Inform. Kew **1915**: 256 (1915). —Obermeyer in F.S.A. **5**(3): 7 (1992). —Newton in Eggli, Illust. Handb. Succ. Pl. **1**: 267 (2001). Type: Illustration t.2634 in Bot. Mag. **53** (1826), cultivated plant of unknown origin.

Stemless plant with creeping rhizome c.2.5 cm thick. Leaves 4–6 per shoot, in rosette, spreading, 40–150 long, including a folded, petiole-like base 7–8 cm long, 3–9 cm wide, lanceolate, irregularly marked with dark green and grey-green transverse bands, margin red. Flowering stem 30–68 cm long; peduncle 12–47 cm long, with 4–5 loose papery bracts, 30–40 × 12–15 mm; rachis 8–20 cm long, raceme densely many-flowered, ± capitate to ovoid, to 21 cm wide. Flowers 1–3 per cluster, greenish-white; floral bracts papery, 1.5–2 cm long; pedicel 3–6 mm long, articulated c.1 mm below flower; perianth 80–130 mm long, tube 65–90 mm long, 2–3 mm wide; lobes 25–35 mm long. Stamens slightly shorter than perianth; style exserted.

Zambia. B: Senanga Dist., 11 km SW of Senanga, 1030 m, fl. 5.viii.1952, *Codd* 7410 (BM, K, PRE). S: Siavonga Dist., c.5 km SW of Chirundu Bridge, fl. 1.ii.1958, *Drummond* 5419 (K, SRGH). **Zimbabwe**. N: Shamva Dist., Shamva, 900 m, fl. 26.xii.1921, *Eyles* 3236 (K, BOL). **Mozambique**. MS: Sussundenga Dist., Dombe, E of Makurupini R., rocky hillside in *Brachystegia* woodland, c.500 m, fl. 12.i.1969, *Bisset* 54 (K, LISC, SRGH).

In Bioko and Congo, Angola and NE Namibia. Woodland on rocky hillsides; 500–1100 m.

Conservation notes: Widespread species; not threatened.

The SABONET national checklists record this species as being also present in Zimbabwe E: Mozambique T: and in Zambia N:, but it has not been possible to confirm this.

Varietal differences are not recognized here. Var. *fernandopoensis* N.E. Br. is only recorded from Bioko.

12. **Sansevieria longistyla** la Croix in Kew Bull. **59**: 617 (2004). Type: Malawi, Salima, sandy lakeshore by Grand Beach Hotel, 480 m, fl. in cult., Kew 23.ii.1976, *Brummitt* 10284 (K).

Stemless herb with thick, creeping orange rhizome c.3 cm thick. Number of leaves per shoot not known. Leaves 33 × 7.3 cm, lanceolate, apiculate, flat but folded at base, dark blue-green, mottled pale blue-green, slightly rough to touch on both upper and lower surfaces, with prominent red margin. Inflorescence 48–58 cm tall; peduncle 22–32 cm long, with 3–5 ovate bracts, 3–3.8 × 1.6–1.8 cm long; rachis 21–27 cm long; raceme 9–12 cm wide, densely many flowered. Flowers 1–3 per bract, suberect, greenish-white (described in *Robson* 1643 as glaucous blue-green at base and apex); pedicels 3–4 mm long, articulated just below flowers; floral bracts papery, 9–10 × 2–3 mm; perianth 80–85 mm long, tube 50–65 mm long, lobes 15–20 mm long. Stigma 100–110 mm long, exserted for 38–40 mm.

Malawi. C: Nkota-Kota Dist., shore of Lake Malawi, between Lake Nyasa and Senga Bay hotels, 480 m, fl. 17.ii.1959, *Robson* 1643 (K).

Not known elsewhere. In sand and thicket near lakeshore; c.500 m.

Conservation notes: Apparently endemic to the shores of Lake Malawi; probably Vulnerable D2.

Pawek 7964 (inflorescence) and 10697 (leaf) from Njakwa Gorge in N Malawi may also belong to this species. The styles are long-exserted but the raceme is much shorter, ± subcapitate, and the flowers are smaller with the perianth 55–60 mm long. *S. downsii*, described by Chahinian from the Njakwa Gorge in 2000, is quite different, with cylindrical leaves and a perianth 33–42 mm long.

13. **Sansevieria metallica** Gérôme & Labroy in Bull. Mus. Hist. Nat. (Paris) **9**: 173 (1903). —N.E. Brown in Bull. Misc. Inform. Kew **1915**: 245 (1915). —Newton in Eggli, Illust. Handb. Succ. Pl. **1**: 267 (2001). Type: Cultivated plant, received at Kew in 1900 from Paris Bot. Garden, fl. iv.1909 (K).

Sansevieria metallica var. nyasica N.E. Br. in Bull. Misc. Inform. Kew **1915**: 247 (1915). Type: Malawi, possibly near Blantyre, coll. 1892, cult. at Kew, 10.ix.1912, *Buchanan* s.n. (K holotype).

Stemless plant with a reddish creeping rhizome. Leaves 1–3 per shoot, ± erect, two bract-like leaves at base to 18 × 3 cm; relatively thin-textured, strongly mottled, broadly lanceolate, narrowed at base for up to 21 cm, 70–110 cm long, 4.5–6 cm wide. Flowering stem 76–105 cm tall; peduncle 45–60 cm long, c.3 cm wide, with several bracts; rachis 31–45 cm long. Flowers in clusters of 2–7, white; pedicel 8 mm long, remnants 5 mm long; perianth tube 13–18 mm long, lobes 10–18 mm long. Fruits not seen.

Malawi. S: Blantyre Dist., 1892, cult. at Kew, 10.ix.1912, *Buchanan* s.n. (K). **Mozambique**. N: Palma Dist., NW of Palma, 10°39'58"S 40°25'13"E, 60 m, 7.xii.2008, *Goyder et al.* 5091 (K, LMA). Z: Morrumbala Dist., 22 km from turn-off Camapo–Nicuadala road to Morrumbala, 100 m, fl. 29.xii.1967, *Torre & Correia* 15788 (LISC). T: Cahora Bassa Dist., 4.5 km from Cabora Bassa dam wall towards Meroeira, 610 m, fl. 7.ii.1973, *Torre, Carvalho & Ladeira* 19033 (LISC). MS: Báruè Dist., 80 km from Catandica (Vila Gouveia) to Changara, 400 m, 31.iii.1966, *Torre & Correia* 15544 (LISC). GI: Inhassoro Dist., Santa Carolina Is., E coast sand flats, st. 3.x.1958, *Mogg* 28792 (J, LISC).

Also in South Africa and possibly Angola. On sandy soils; sea level–900 m.

Conservation notes: Probably Lower Risk near threatened.

The species bears some resemblance to *S. hyacinthoides*, but has larger leaves and usually smaller flowers. There are also only 1–3 leaves per growth (often only 2), whereas *S. hyacinthoides* has leaves in a rosette.

Varietal differences are not recognized here. Var. *metallica* is known only from KwaZulu-Natal in South Africa, while var. *longituba* N.E. Br. (in Bull. Misc. Inform. Kew **1915**: 247, 1915) is said to have flowers with the tube and lobes each 29–30 mm long. The type is a cultivated plant with the origin given as Tropical Africa.

14. **Sansevieria pearsonii** N.E. Br. in Bull. Misc. Inform. Kew **1911**: 97 (1911); in Bull. Misc. Inform. Kew **1915**: 216 (1915). —Obermeyer in F.S.A. **5**(3): 8 (1992). Type: S Angola, km 108.5 on Mossamedes railway, 500 m, 30.iv.1909, *Pearson* 2073 (K holotype). FIGURE 13.2.7.

Sansevieria deserti N.E. Br. in Bull. Misc. Inform. Kew **1915**: 208 (1915). Types: Botswana, Ngamiland, edge of Boteti R., 25.vii.1897, *E.J. Lugard* 9 (K syntype) & shores of Lake Ngami, iv.1890, *Nicholls* s.n. (K syntype).

Sansevieria rhodesiana N.E. Br. in Bull. Misc. Inform. Kew **1915**: 212 (1915). Types: N Botswana, uncertain locality, cultivated at Kew, 27.xi.1913, as Imperial Inst. 50998 (K syntype); & Zimbabwe, 1908 (fl. i.1912), as Kongye Difa from Imperial Inst. 677/1908 (K syntype).

Stemless plant with woody, creeping, bright orange-red rhizome 1.5–2.5 cm thick. Leaves in 2 rows, bases overlapping, 4–7 per shoot, 57–150 cm long, 1–2.5 cm in diameter, ± cylindrical, ridged, channelled towards base (occasionally extending to leaf tip), edges of channel acute,

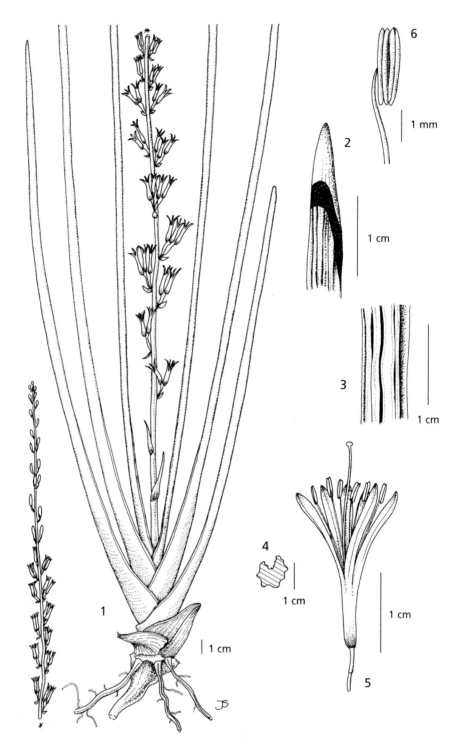

Fig. 13.2.**7**. SANSEVIERIA PEARSONII. 1, habit; 2, leaf apex; 3, part of leaf showing vertical ridges; 4, cross-section of inner leaf, 150 cm from base; 5, flower; 6, anther. All from *Fanshawe* 6946. Drawn by Judi Stone.

hardened, red-brown in lower part, green further up, apex acute, hard. Flowering stem 58–90 cm long; peduncle 10–17 cm long, with 3 bracts to c.45 cm long. Flowers green, cream, buff-pink or yellow, sweetly scented, in clusters of 4–7, set c.1 cm apart; pedicel 4–6 mm long, slender, remnants 2–4 mm long; perianth 18–20 mm long, tube 7–13 mm long, lobes 7–12 mm long. Stamens similar length to perianth; style slightly exserted. Fruit 7 mm diameter.

Botswana. N: Ngamiland Dist., banks of Boteti (Botletle) R., fl. 25.vii.1897, *Lugard* 9 (K). **Zambia**. B: Sesheke Dist., fl. 18.vii.1962, *Fanshawe* 6946 (K, NDO). W: Kitwe, fl. 6.vii.1955, *Fanshawe* 2360 (K, NDO). C: Kafue Dist., Kafue, 1060 m, st. xi.1932, *Trapnell* 1135 (K). S: Katombora Dist., Katombora, c.900 m. st. 26.viii.1947, *Greenway & Brenan* 7990 (K). **Zimbabwe**. N: Binga Dist., Mwenda Research Station, fl. 5.vi.1966, *Grosvenor* 121 (K, SRGH). W: Hwange Dist., 24 km NE of Sebungwe–Zambezi confluence, fl. 11–16.v.1956, *Plowes* 1989 (K, SRGH). C: Harare Dist., near Hunyani R., 1360 m, fl. 15.ix.1946, *Wild* 1240 (K, SRGH). E: Mutare Dist., hill overlooking Dora R., Quagga's Hoek commonage, 1100 m, fl. 12.vii.1953, *Chase* 5011 (BM, SRGH). S: Zvishavane Dist., Zvishavane (Shabani) to Masvingo (Fort Victoria), 1500 m, fl. 9.vii.1955, *Plowes* 1863 (K, SRGH). **Mozambique**. Z: Gilé Dist., Gilé, c.10 km from Mt Gilé, c.300 m, fr. 21.xii.1967, *Torre & Correia* 16693 (LISC). T: Chifunde Dist., between Régulo Bene and Furancungo, 23.4 km from Régulo Bene, fl. 14.vii.1949, *Barbosa & Carvalho* 3602 (K, LISC, LMA). MS: Caia Dist., Sena, fr. 24.x.1906, *Johnson* 12 (K). GI: Govuro Dist., Nova Mambone (Mambone), fl. 7.viii.1907, *Johnson* 268 (K) [possibly a cultivated plant originally from Sena]; Massinga Dist., Funhalouro, 15.v.1941, *Torre* 2703b (LISC).

Also in S Tanzania, Angola, Namibia and N South Africa. Rocky hillsides, sometimes under trees or on termitaria on poor sandy soils, often forming dense colonies; 100–1500 m.

Conservation notes: Widespread species; not threatened.

Reported as cultivated in gardens in Zimbabwe by Biegel (Checklist. Orn. Pl. Rhod. Gdns., Rhod. Agric. J. Res. Report **3**: 96, 1977).

S. sordida N.E. Br. from Kenya differs from *S. pearsonii* in having leaves that are rough to the touch, rather than smooth.

15. **Sansevieria pedicellata** la Croix in Kew Bull. **59**: 620 (2004). —Rulkens & Baptista in Sansevieria **20**: 2 (2009). Type: Mozambique, Sussundenga Dist., Musapa Gap, 1.ii.1962, *Wild* 5638 (K holotype).

Stemless plant with creeping rhizome 1.5–5 cm thick. Number of leaves per growth 4–6(15). Leaves 72 × 6 cm, linear, plain green with no spots or banding, red line around margin and a thin, white line outside that, red margin apparently folded where leaf becomes narrower; apical 5 cm narrowed abruptly to a sharp point. Peduncle 30 cm long, c.4 mm wide, with 4 broadly lanceolate, acute bracts 4–6 cm long, 1.4 cm wide; rachis over 27 cm long (broken in specimen seen). Flowers whitish, in clusters of 4–6, set c.1 cm apart; pedicel remnants 15–18 mm long, deciduous part 1–2 mm long; perianth 90–100 mm long; tube 80–90 mm long, lobes 17–20 mm long. Fruits not seen.

Mozambique. MS: Sussendenga Dist., Musapa Gap, c.3 km from Zimbabwe border, fl. 1.ii.1962, *Wild* 5638 (K).

Only known from Manica Province. Among rocks by streams; 700–1500 m.

Conservation notes: Endemic to the Chimanimani Mts and Manica Province in Mozambique. Probably Vulnerable D2 or Endangered.

Rulkens & Baptista (2009) report that *S. pedicellata* is found elsewhere in Manica Province at 700–1000 m, near Chimoio town, Mt Zembe in the east, Mt Gurungue and Chitsamba in the west bordering Zimbabwe, Mt Guro and Chimane. It appears to be the same taxon as *Sansevieria* "Chimanimani Mountains" and *S.* "Chipinge" in Chahinian (2005). Reportedly moth pollinated.

The only material available of this species is the type specimen, but the very long pedicel remnants and the elongated, spike-like tip of the leaves are so distinctive that it was not possible to assign it to any other species.

16. **Sansevieria scimitariformis** D.J. Richards in Sansevieria **5**: 8 (2002). Type: Zimbabwe, Shamva, rocky outcrop 9 km on road to Makuruanopamaenza Hills, from cultivated plant, 24.v.1993, *Richards* R2259 (SRGH holotype).

Stemless plant forming dense clumps, with a creeping rhizome 3–4 cm thick. Leaves 4–6 per shoot, 50–70 × 6–10 cm, up to 3 cm thick near base, rigid, recurved, dark green, with shallow longitudinal grooves on both upper and lower surfaces, sometimes transverse bands of whitish patches on both surfaces; margins with a narrow brown line and thin, fibrous strips on extreme edge; apex blunt. Peduncle c.15 × 2 cm, with 4 bracts clasping stem, up to 4 × 4 cm. Inflorescence rachis c.10 cm long, inflorescence a dense, ± capitate head. Flowers pale green, in clusters of 4–6; pedicels 5–6 mm long; bracts 20 mm long; perianth tube 100–120 mm long, purplish, lobes whitish. Style exserted by c.40 mm; stamens exserted 10–15 mm. Fruit globose, orange-red when mature, surface wrinkled.

Zimbabwe. N: Shamva Dist., 3–4 km N of 9 km peg on Shamva– Mutawatawa road, 24.v.1993, *Richards* R2259 (SRGH); Darwin Dist., 37.5 km along road from Mt Darwin to Mavuradonha Mts, *Richards* R998 (SRGH).
Not known elsewhere. Rocky outcrops; c.1000 m.
Conservation notes: Apparently endemic to middle reaches of the Mazoe valley; Data Deficient, probably Vulnerable D2.
This species resembles *S. hallii* Chahin. but has more leaves per growth, the leaves are wider and the inflorescence is held well above ground level.
Trapnell 1233 from Zambia C: (Chisamba R. on Kaoma (Mankoya) border, 1030 m, st. v.1933) may belong to this species. A drawing shows a growth of about 6 leaves with a note saying they are smooth above and scabrid below.

17. **Sansevieria sinus-simiorum** Chahin. in Sansevieria **3**: 24 (2002). Type: Malawi, Mangochi Dist., c.2 km S of Monkey Bay, 1985, *Burdett* s.n. in *Chahinian* 316 (MO holotype, NYBG).

Stemless perennial; rhizome short, c.50 mm thick. Leaves 8–10 per shoot, crowded, to 100 cm long, 6 cm wide at base, gradually tapering to an obtuse apex, straight or recurved, waxy dark green with some obscure cross-banding when mature, a round channel about $1/3$ of leaf width at base, widening to ± full leaf width at top, numerous vertical lines in channel and on back of leaf, edges of channel acute, edged with withered fibres. Peduncle c.30 cm tall, with c.6 bracts 32–35 mm long, withered by flowering time; inflorescence a many-flowered, capitate raceme, c.19 cm wide; pedicel 6 mm long. Flowers white, tinged with green, scented at night; perianth tube 82–84 mm long, slightly inflated to 4 mm wide in middle, lobes 19–21 mm long, 3 mm wide. Style c.126 mm long, filiform, c.16 mm longer than stamens. Filaments 28–32 mm long, anthers 3–4 mm long, exserted.

Malawi. N: Chitipa Dist., Chambo Mission, 4 km N of Chisenga, 1500 m, fr. 3.iii.1982, *Brummitt, Polhill & Banda* 16307 (K). C: Mangochi Dist., c.2 km S of Monkey Bay, rocky face overlooking Lake Malawi, 1985, fl. in cult., *Burdett* in *Chahinian* 316 (MO, NYBG).
Not known elsewhere. On rock faces; 480–1500 m.
Conservation notes: Apparently endemic to Malawi; probably Endangered.

18. **Sansevieria stuckyi** God.-Leb., Sansev. Gigant. Afr. Orient: 13 (1901) ex Gérôme in Bull. Mus. Hist. Nat. (Paris) **9**: 173 (1903). —N.E. Brown in Bull. Misc. Inform. Kew **1915**: 219 (1915). —Newton in Eggli, Illust. Handb. Succ. Pl. **1**: 270

(2001). —Mbugua in F.T.E.A., Dracaenaceae: 33 (2007). Type: Mozambique, Zambézia, Boror, 27.xi.1901, *Stucky* s.n. (K holotype).

Stemless plants forming large clumps. Rhizome creeping, 2–2.5 cm thick, usually with 1 leaf per shoot, occasionally 2, rarely 3. Leaves erect, up to 2.8 m tall, 5 cm wide at base, dark green with whitish transverse lines when alive, smooth to the touch, cylindrical with a shallow channel extending from base to apex; apex acute. Flowering stem 20–43 cm long, peduncle purplish marked with green, 2–3 scarious bracts to 45 mm long; rachis 7 cm long, inflorescence capitate, c.34 cm wide, densely many-flowered. Flowers white, sometimes tinged with purple; tube 90–100 mm long, lobes 40 mm long. Stamens and style well exserted.

Zimbabwe. E: Burma or Honde valley, 20.vii.1987, *Richards* 54 (SRGH). **Mozambique**. N: Malema Dist., c.40 km from Malema (Entre Rios) to Ribáuè, Serra Murripa, c.1000 m, in bud 14.xii.1967, *Torre & Correia* 16510 (LISC). Z: Mocuba Dist., Boror, st. 27.xi.1901, *Stucky* s.n. (K).

Also in S Kenya. Probably dry bush; 50–1000 m.

Conservation notes: Data Deficient, but probably Vulnerable D2 within the Flora area.

19. **Sansevieria subspicata** Baker in Gard. Chron., ser. 3 **6**: 436, t.115 (1889). —N.E. Brown in Bull. Misc. Inform. Kew **1915**: 234 (1915). —Newton in Eggli, Illust. Handb. Succ. Pl. **1**: 271 (2001). Type: Mozambique, Maputo (Delagoa Bay), 1866, *Monteiro* s.n. (K holotype).

> *Sansevieria subspicata* Baker var. *subspicata* sensu Mbugua in F.T.E.A., Dracaenaceae: 17 (2007).

Stemless, rhizomatous plant; rhizome 1–2.5 cm thick. Leaves 4–10 per shoot, in a rosette, ± erect, blade lanceolate to oblanceolate, acute, flat, relatively thin-textured, 18–38 × 2.8–7 cm, narrowing to a channelled petiole 5–33 cm long, sometimes as long as the blade. Flowering stem 30–60 cm tall; peduncle 15–30 cm long, with 4 acuminate bracts 3–5.5 cm long; rachis similar length to peduncle, rather laxly many-flowered; pedicel 3–5 mm long, articulated c.1 mm below flower. Flowers greenish-white, 1–2(4) per cluster; bracts c.4 mm long; perianth 32–48 mm long, tube 20–30 mm long, lobes 12–25 mm long. Fruit 1–3-lobed, lobes globose, c.8 mm wide.

Mozambique. N: Mogincual Dist., Quinga, 5 m, 28.ii.1964, *Torre & Paiva* 11441 (LISC). Z: Maganja da Costa, Gobene Forest, c.20 m, fl. 12.ii.1966, *Torre & Correia* 14539 (LISC). MS: Sussundenga Dist., Dondo, fr. 6.v.1942, *Torre* 4076 (LISC). GI: Massinga Dist., Pomene, c.1 km past airstrip, st. 24.ix.1980, *Jansen, de Koning & Zunguze* 7499 (K). M: Maputo, margin of Rio Matola, near bridge, fl. 18.ii.1961, *Balsinhas* 279 (K, LISC).

Not known elsewhere. Scrub and thickets on coastal sandy soils; sea level–50 m.

Conservation notes: Endemic to coastal Mozambique; Lower Risk near threatened.

ARECACEAE (PALMAE)

by J. Dransfield

Small to large, single-stemmed or clustered, armed or unarmed, monoecious or dioecious plants, sometimes dying after flowering. Stems woody, slender to massive, usually unbranched, rarely branching dichotomously, lacking cambium but sometimes increasing in diameter by diffuse growth, roots adventitious, sometimes modified into spines. Leaves alternate, spirally arranged, rarely distichous or tristichous; sheath initially always tubular at base, later frequently splitting, unarmed or armed with spines, sometimes forming a crown-shaft; petiole usually present, terete, or variously channelled or ridged below, unarmed or with spines; hastulae usually present in palmate leaves; blade palmate, costapalmate, pinnate, bipinnate, or bifid, or entire and pinnately veined, splitting along the adaxial or abaxial folds, rarely splitting between folds or not splitting; segments or leaflets lanceolate or linear to rhomboid or wedge-shaped, V-shaped (induplicate) or ʌ-shaped (reduplicate), single-fold or composed of many folds, tips acute, acuminate, truncate, oblique or bifid, or irregularly toothed or lobed, sometimes armed with spines or bristles along the margins and/or main veins, transverse veinlets conspicuous or obscure; proximal leaflets sometimes modified as spines (acanthophylls), rachis prolonged distally into a climbing whip (cirrus) in many climbing palms, sometimes also bearing acanthophylls. Inflorescences axillary, infrafoliar, interfoliar, or aggregated into a suprafoliar compound inflorescence, spicate or branched up to 6 orders, in some species of *Calamus* inflorescence modified as a climbing whip (flagellum); peduncle short to long; prophyll usually 2-keeled; peduncular bracts 0 to many; rachis shorter or longer than the peduncle; rachis bracts similar to peduncular bracts, or dissimilar, or much reduced; rachillae (flower-bearing branches) short to long, slender to massive, rachilla bracts conspicuous to minute or apparently lacking, sometimes forming pits containing the flowers. Flowers hermaphrodite or unisexual, then similar or dimorphic, sessile or stalked, borne singly or in cincinni; perianth usually clearly differentiated into sepals and petals; sepals (2)3 (rarely more); petals (2)3 (rarely more); stamens (3)6 (or 950 or more), staminodes ranging from toothlike to well developed, rarely absent; gynoecium apocarpous with (1)3(4) carpels, or variously syncarpous with 3 or rarely more (to 10) locules, or pseudomonomerous with 1 fertile locule, carpels glabrous, variously hairy, or covered with imbricate scales, styles distinct or connate or not clearly differentiated, stigmas erect or recurved; ovule solitary in each locule, pistillode present or absent in the staminate flower. Fruit usually 1-seeded, sometimes 2–10-seeded, ranging from small to very large, stigmatic remains apical, lateral, or basal; epicarp smooth or hairy, prickly, corky-warted or covered with imbricate scales, mesocarp fleshy, fibrous or dry, endocarp not differentiated or thin, or thick and then often with 3 or more pores at, below, or above the middle. Seed with thin or sometimes fleshy testa (sarcotesta), endosperm homogeneous or ruminate; embryo apical, lateral, or basal. Germination adjacent or remote; seedling leaf simple and entire, bifid, palmate, or pinnate.

A family of 183 genera and c.2500 species, distributed widely in the tropics and subtropics of both hemispheres.

Genera Palmarum (Dransfield, Uhl, Asmussen, Baker, Harley & Lewis, Genera Palmarum, ed.2: 2008) provides an introduction to the family and its classification. The order of genera used here follows this publication.

The palms are economically very significant, being the third most important family after grasses and legumes. They include commercial crops such as the African Oil Palm, the Date Palm and the Coconut, but there are many species of great significance to life in rural areas in the tropics and subtropics, e.g. *Borassus aethiopum*, *Raphia farinifera* and *Hyphaene* species in the Flora area. Their uses range from food, biofuel, construction material, handicrafts, medicine, personal adornment and horticulture.

Cultivated palms

We have little information on the palms commonly cultivated in the Flora area. The number of species cultivated as ornamentals seems to increase each year as more species are introduced from the wild. The following have been recorded by Biegel (Checklist Ornam. Pl. Rhod. Gdns., Rhod. Agric. J. Res. Report **3**: 1977), by Maroyi (Kirkia **18**: 186, 2006), and by T. Müller and M. Bingham (pers. comm.). However, many palms that are widespread elsewhere in cultivation are missing. Useful illustrated sources of information are: R.L. Riffle & P. Craft, Encyclopedia of Cultivated Palms (2003), Fairchild On-line Guide to Palms (www.palmguide.org) and Palmweb (www.palmweb.org).

Archontophoenix alexandrae (F. Müll.) H. Wendl. & Drude, Alexandra or King Palm, native to Australia. A medium-sized single-stemmed monoecious palm with a crown of c.10 pinnate leaves, neatly abscissing to give a clean stem well-marked with leaf scars, leaf sheaths forming a distinct crown-shaft; leaflets acute, regularly arranged, paler beneath. Inflorescence 50–100 cm long, held below leaves, branched to 4 orders, gleaming pale cream-coloured branches with cream-coloured flowers. Fruits globose, bright red, c.10 mm diameter. Often planted in gardens in Zimbabwe (Harare) and Zambia (Lusaka).

Archontophoenix cunninghamiana (H. Wendl.) H. Wendl. & Drude, Piccabeen or Bangalow Palm, native to Australia. Similar to *A. alexandrae* but differing in leaflets being same colour above and below, and with lilac-coloured flowers. Commonly planted in gardens in Zimbabwe (Harare) and Zambia (Lusaka).

Caryota urens L., Fishtail or Kitul Palm, native to S India and Sri Lanka. Robust single-stemmed monoecious palm with stems to 15 m tall. Leaves bipinnate with fish-tailed leaflets. Stems producing inflorescences in a basipetal sequence, stem dying after flowering and fruiting. Inflorescence pendulous, to 1 m long. Fruits ripening reddish, c.12 mm diameter, filled with irritant needle crystals. Occasionally planted in Zambia (Lusaka), Zimbabwe (Harare) and Mozambique (Maputo).

The smaller multistemmed *Caryota mitis* Lour. is also said to be present in Zambia.

Chamaedorea elegans Mart. (= *Collina elegans* (Mart.) Liebm.), Parlour or Neanthe Palm, native to Mexico. Dwarf, single-stemmed dioecious palm to 40 cm, often used in gardens for shade bedding. Leaves pinnate, to 40 cm long, neatly abscissing from stem to leave numerous close leaf scars; leaflets acuminate, regularly arranged, bright green. Inflorescence erect, held between leaves, numerous slender branches with minute globular yellow flowers, scented of blood. Fruit ± rounded, c.7 mm diameter. A common pot plant, also planted in gardens, in Zimbabwe (Harare), and also in Zambia (Lusaka). Other species of *Chamaedora* are increasingly being planted in Harare.

Dypsis lutescens (H. Wendl.) Beentje & J. Dransf. (= *Chrysalidocarpus lutescens* H. Wendl.), Golden Cane, Areca or Butterfly Palm, native to coastal E Madagascar. Densely clustering monoecious palm with green stems to 4 m tall. Leaves pinnate, neatly abscissing to produce a clean stem with conspicuous nodal scars, leaf sheaths forming a distinct, often grey or yellowish crown-shaft; petioles often golden-coloured, leaflets regularly arranged, curved, tips acuminate. Inflorescence held between or below leaves, branched to 3 orders, with numerous yellowish branches. Fruits yellowish, ellipsoid, to 18 × 11 mm. Often planted in Zimbabwe (Harare).

Howea fosteriana (C. Moore & F. Müll.) Becc., Kentia Palm, native to Lord Howe Is. Single-stemmed monoecious palm to 8 m tall, usually less. Leaves pinnate, neatly abscissing; sheaths with fibrous margins, not forming a crown-shaft; leaflets acute, regularly arranged, somewhat curved. Inflorescence unbranched, held between leaves, several emerging from the same leaf axil. Flowers in deep pits on the spike. Fruit dull reddish brown, ellipsoid, to 4 cm long. Occasionally seen as a pot or garden plant in Zimbabwe (Harare).

Hyophorbe verschaffeltii H. Wendl., Bottle Palm, native to Rodrigues (Mascarene Is.). Single-stemmed palm with a grossly swollen, bottle-shaped trunk. Leaves pinnate, neatly abscissing to produce distinct nodal scars; leaf sheaths forming a distinct crown-shaft; leaflets acute, irregularly grouped along rachis, held in different planes giving leaf a plumose appearance. Inflorescence held below leaves, branched to 4 orders, numerous branches with minute flowers held in vertical rows. Fruit ellipsoid, black, c.10 × 6 mm. Occasionally seen as a pot or garden plant in Zimbabwe (Harare) and in Zambia (Lusaka).

Hyophorbe lagenicaulis (L.H. Bailey) H.E. Moore is very similar but has regularly arranged leaflets.

Livistona chinensis (Jacq.) R. Br., Chinese Fan Palm, native to S China and Vietnam. Moderately robust, single-stemmed hermaphrodite palm. Leaves costapalmate with neatly pendulous segments, leaves not neatly abscissing; leaf bases not cleft, with abundant fibrous network. Inflorescence held between leaves, to 2 m long, branched to 4 orders. Flowers held in groups up to 5, small, pale green, hermaphrodite. Fruit subglobose to ellipsoid, shiny, bluish-green, to 20 × 12 mm. Planted in Mozambique (Maputo) and occasionally seen as a container plant in Zimbabwe (Harare).

Lytocaryum weddellianum (H. Wendl.) Toledo (= *Syagrus weddelliana* (H. Wendl.) Becc.), native to S Brazil. Dwarf single-stemmed monoecious palm. Leaves pinnate, not neatly abscissing; leaf sheaths fibrous, not forming a crown-shaft; leaflets acute, very regularly arranged, mid to dark green above, gleaming white beneath. Inflorescence held between leaves, sparsely branched to 1 order. Fruit globose, yellowish at maturity, epicarp and mesocarp splitting to display smooth endocarp with 3 pores, like a miniature coconut. Occasionally seen as a pot plant in Zimbabwe (Harare).

Phoenix canariensis Chabaud, Canary Island Date Palm, native to the Canary Islands. Massive single-stemmed dioecious palm. Leaves massive, not neatly abscissing, trunk eventually with very close leaf scars; leaf sheaths fibrous; leaflets acute, held in different planes, basal ones modified as spines. Inflorescence held between leaves, branched to 1 order, with numerous cream-coloured flowers. Fruit orange, like a miniature date, up to 20 mm long. Commonly planted in Zimbabwe (Bulawayo and Harare) and in Zambia (Lusaka).

Phoenix roebelenii O'Brien, Pygmy Date Palm, native to S China and Laos. Usually a single-stemmed dainty dioecious palm, rarely exceeding 2 m tall. Leaves not neatly abscissing, trunk eventually with close leaf scars; leaf sheaths fibrous; leaves usually 1–1.5 m long, with very regularly arranged, soft curved leaflets. Inflorescence held between leaves, branched to 1 order. Fruit ellipsoid, black, c.8 mm long. Occasionally seen as a pot or garden plant in Zimbabwe (Harare).

Ptychosperma elegans (R. Br.) Blume, Solitaire Palm, native to Australia. Moderate-sized single-stemmed monoecious palm, usually not exceeding 5 m tall. Leaves pinnate, usually about 10 in crown, neatly abscissing to leave a clean trunk with conspicuous leaf scars; leaf sheaths forming a green crown-shaft; leaflets regularly arranged, ending abruptly with an oblique tattered apex. Inflorescence held below leaves, branched to 3 orders, with spreading greenish branches. Fruit bright red, c.15 × 10 mm. Occasionally planted in Zimbabwe (Harare).

Rhapis excelsa (Thunb.) Henry, Partridge Cane Palm, native to S China. Densely clustered dioecious palm with short stems to 2 m tall. Leaves palmate, blade divided between major folds, not neatly abscissing, stem covered in a network of dark fibrous sheaths; leaf blade 80 cm wide or less. Inflorescence between leaves branched to 3 orders. Fruit very rarely produced in cultivation, c.10 mm diameter. A popular pot plant and also planted in gardens in Zimbabwe (Harare).

Roystonea regia (Kunth) O.F. Cook, Cuban Royal Palm, native to Cuba and S Florida. Immense single-stemmed monoecious palm. Leaves pinnate, neatly abscissing to produce a clean trunk; leaf sheaths forming a massive green crown-shaft to 2 m long; leaflets acuminate, held in many planes, giving the leaf a plumose appearance. Inflorescence massive, held below leaves, branched to 4 orders or more. Fruit purplish black, ellipsoid, 9–15 × 7–11 mm. Planted in Mozambique (Maputo) and occasionally in Zimbabwe (Harare).

Roystonea oleracea (Jacq.) O.F. Cook, Caribbean Royal or Cabbage Palm, native to the Caribbean islands. Very similar to *R. regia* but tending to be even larger, with a hemispherical rather than globose crown, oldest leaves tend to be held horizontally. Occasionally planted in Zimbabwe (Harare).

Syagrus romanzoffiana (Cham.) Glassman (= *Arecastrum romanzoffianum* (Cham.) Becc.), Queen Palm, native to S Brazil and Argentina. Robust, single stemmed monoecious palm with trunks to 12 m tall. Leaves pinnate, not neatly abscissing, trunk obscured by old leaf bases; leaf sheaths very fibrous, not forming a crown-shaft; leaflets numerous, acute or slightly bilobed, irregularly arranged and held in different planes giving the leaf a plumose appearance. Inflorescence held between leaves, branched to 1 order, with an inconspicuous prophyll and a large woody, grooved, hooded peduncular bract. Flowers golden yellow. Fruit ovoid, orange, c.2.5 mm diameter, endocarp with 3 pores like a coconut, rough internally with irregular intrusions. Commonly planted in Zimbabwe (Harare) and also recorded from Zambia (Lusaka).

Trachycarpus fortunei (Hook.) H. Wendl., Chinese Windmill Palm, native to China. Moderate-sized single-stemmed dioecious palm. Leaves palmate, not neatly abscissing, leaf sheaths with abundant dark brown fibres; leaf blade irregularly divided by shallow and deep splits. Inflorescence held between leaves, branched to 3 or 4 orders, with conspicuous inflated bracts and crowded golden flowers. Fruit bluish-black, kidney-shaped, c.10 mm wide. Recorded from Zimbabwe (Harare).

Washingtonia filifera (André) de Bary, Desert Fan Palm, native to California and adjacent N Mexico. Robust single-stemmed hermaphrodite palm. Leaves costapalmate, not neatly abscissing, old leaves forming a conspicuous skirt of dead leaves (often maliciously burned or pruned away); leaf bases fibrous, with very conspicuous triangular cleft below petiole; leaf blade with many irregularly pendulous segments with fibrous margins. Inflorescence held between leaves, pendulous to 3 m or more long, with woody sword shaped bracts. Fruit black, ellipsoid, c.9 × 5 mm. Recorded from Zimbabwe (Harare).

Washingtonia robusta H. Wendl., Desert Fan Palm, native to N Mexico. Very similar and more common than *W. filifera*, but has a taller and thinner trunk, a bright green rather than ± glaucous leaves held in a tighter crown, and fewer fibres on segment margins. Commonly planted in Zimbabwe (Harare) and also recorded from Zambia (Lusaka).

1. Leaves palmate or costapalmate 2
– Leaves pinnate 3
2. Petioles armed with irregular black teeth, varying greatly in size and shape; leaf segments with conspicuous cross-veins and lacking black spots; fruit ± rounded, containing 1–3 separate pyrenes, with apical stigmatic remains . . . . **6. Borassus**
– Petioles armed with regular large upward pointing black teeth; leaf segments lacking conspicuous cross-veins, but with abundant black punctiform scales; fruit varied in shape, 1-seeded, if more developing, then fruit lobed, each lobe with 1 seed, stigmatic remains basal . . . . **5. Hyphaene**

3. Climbing palms 4
 - Erect tree palms 5
4. Leaf sheaths unarmed; climbing whip borne at tip of leaf rachis (cirrus), armed
   with large mostly paired, reflexed acanthophylls (modified leaflets) . . . .
   . . . . **1. Eremospatha**
 - Leaf sheaths densely spiny; climbing whip emerging from leaf sheath not leaf tip
   . . . . **3. Calamus**
5. Unarmed single-stemmed palm; fruits large, rounded, at least 10 cm diameter
   . . . . **7. Cocos**
 - Palms variously armed on petioles or leaflets, fruit smaller than 10 cm diameter,
   if this wide then ± fusiform 6
6. Massive palms, leaflets bearing abundant spines on upper surface of midrib and
   on margins; fruit covered with large shiny brown reflexed scales . . . . **2. Raphia**
 - Robust palms; leaflets unarmed; fruit not covered with reflexed scales 7
7. Leaflets V-shaped in cross-section (induplicate), basalmost leaflets modified as
   spines; petiole lacking marginal teeth; dioecious; inflorescence bearing a
   prophyll, peduncular bract absent; rachillae (flower-bearing branches) becoming
   widely spreading; fruit without a 3-pored stone . . . . **4. Phoenix**
 - Leaflets Λ-shaped in cross section (reduplicate), basalmost leaflets eroding
   leaving spiniform midribs; petiole base armed with sharp teeth; monoecious but
   inflorescence mostly unisexual; rachillae very crowded and congested among
   leaf sheaths; fruit with a conspicuous 3-pored stone . . . . **8. Elaeis**

1. **EREMOSPATHA** (G. Mann & H. Wendl.) G. Mann & H. Wendl.

Eremospatha (G. Mann & H. Wendl.) G. Mann & H. Wendl. in Kerchove, Les
Palmiers: 224 (1878). —Beccari in Webbia **3**: 270–294 (1910). —Dransfield *et al.*,
Genera Palmarum, ed.2: 150 (2008).

Clustering, high-climbing, spiny, hermaphrodite palms, not dying on flowering; stems suckering sympodially. Leaves displaying plasticity of form from flabellate juvenile leaves to pinnate
adult leaves, adult and some juvenile stems terminating in a cirrus armed with reflexed thorns
and more massive paired reflexed acanthophylls; leaf sheath unarmed, usually glabrous; ocrea
well developed, usually persistent, unarmed, truncate; petiole present in juvenile foliage, very
short to absent in adult leaves; rachis usually with lateral reflexed spines; leaflets few to many,
basal few often modified as spiny 'aphlebiae', reflexed and clasping stem (function not known);
distal leaves usually armed with marginal spines, other leaves unarmed, lanceolate, suborbicu-
lar to rhomboid, occasionally apiculate, normally part of distal margins praemorse, occasionally
with scattered brown scales. Inflorescence axillary, shorter than leaves, emerging from ocrea,
branching to 1 order; peduncle short; rachillae arranged subdistichously, spreading, subtended
by very small inconspicuous bracts with triangular limbs, with paired flowers over their length,
each pair subtended by a minute annular bract. Flowers hermaphrodite; calyx tubular with 3
low triangular lobes; corolla ± twice calyx length, with 3 short triangular lobes. Stamens 6, borne
near mouth of corolla-tube, united basally, free filaments very short, each bearing a small pair
of anthers. Ovary covered with reflexed scales. Fruit 1–3-seeded, covered with often brightly
coloured reflexed scales; mesocarp fleshy, endocarp not differentiated. Seed sub-basally
attached, usually ellipsoidal, sometimes slightly lobed.

10 species in humid tropical Africa, only one in southern Africa.
A full account of the genus has recently been published by Sunderland (Field
Guide Rattans Africa, 2007).

Eremospatha cuspidata (G. Mann & H. Wendl.) H. Wendl. in Kerchove, Les
Palmiers: 244 (1878). —Wright in F.T.A. **8**: 112 (1901). —Tuley in Palms Africa:

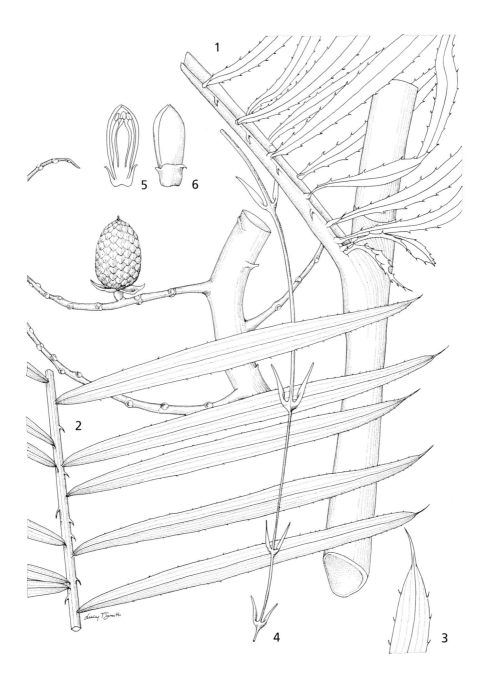

Fig. 13.2.**8**. EREMOSPATHA CUSPIDATA. 1, stem (× ²/₃); 2, leaflets (× ⁴/₉); 3, leaflet tip (× 1); 4, acanthophylls (× ²/₃); 5, flower (× 2²/₃); 6, fruit and infructescence (× 1¹/₃). 1–3,6 from *Sunderland* 1909, 4 from *Sunderland* 1792, 5–6 from *Sunderland* 1922. Drawn by Lucy Smith. From Sunderland, Field Guide Rattans of Africa (2007).

49 (1995). —Sunderland, Field Guide Rattans Africa: 30 (2007). Type: Gabon, Ogooué R., vii.1861, *Mann* 1043 (K holotype). FIGURE 13.2.**8**.

Calamus cuspidatus G. Mann & H. Wendl. in Trans. Linn. Soc. **24**: 434 (1864).

Clustering slender palm climbing to 12–15 m. Stems circular in cross-section, without sheaths, 10–15 mm in diameter; internodes 11–15 cm long. Leaf sheath longitudinally striate, sparsely covered with brown-black indumentum; ocrea obliquely truncate, extending c.1 cm above rachis; knee absent. Leaves sessile, up to 2 m long; rachis 1–1.3 m long, flattened on upper surface, rounded below, becoming trapezoid then rounded in cross-section distally, armed along margins with reflexed, bulbous-based spines, indumentum absent; cirrus 50–75 cm long, unarmed. Leaflets up to 15–20 on each side of rachis, linear-lanceolate, 22–30 × 1.6–2(3) cm broad at widest point, abruptly contracted at base, a fine 0.8–1.2 cm long (rarely 3 cm) apiculum at apex, discolorous, armed along margins with abrupt black-tipped spines, 5–7 moderately conspicuous transverse veinlets 2–3 mm apart; lowermost leaflets smaller, linear-ovate, reflexed and laxly swept back across stem; acanthophylls in pairs c.3 cm long, at 45° to cirrus. Inflorescence glabrous, 30–38 cm long, rarely less than 20 cm; peduncle 10–18 cm long; rachis 20–30 cm long, erect or horizontal; rachillae distichous, 10–12 on each side, 5–12 cm long, decreasing distally; rachis bracts acuminate, less than 3 mm long, attached to inflorescence axis for 1–1.2 cm; flower cluster subtended by c.1 mm long incomplete bracts. Flowers in close pairs, sweetly-scented at anthesis; calyx 4–5 × 6 mm wide at mouth, shallowly lobed; corolla 0.7–1 × 0.4–0.5 mm, divided to $^1/_4$ its length; stamens united into a 4–6 mm long epipetalous ring, free filaments c.0.5 mm, anthers c.1 mm. Ovary 3–4 × 2–2.5 mm, tipped by c.2.5 mm long style. Fruit 1-seeded, ± cylindrical, 2–2.4 × 1.6–2 cm with 18–21 vertical rows of scales. Seed compressed, 1.6–2 × 0.8–1 × 0.6–0.8 cm deep, flattened on one side with a shallow linear depression.

Zambia. W: Mwinilunga Dist., Lisombo R., 12.vi.1963, *Loveridge* 931 (K, SRGH).

Also in Angola; widespread in humid Equatorial Africa. Fringes of swamp forests (mushitu); c.1200 m.

Conservation notes: Only known from one locality in the Flora area, where it is probably Vulnerable; but Least Concern globally.

2. **RAPHIA** P. Beauv.

Raphia P. Beauv., Fl. Owar. **1**: 75 (1809). —Beccari in Webbia **3**: 37–130 (1910). —Chevalier in Rev. Bot. Appl. **12**: 198–213 (1932). —Russel in Kew Bull. **19**: 173–196 (1965). —Otedoh in J. Nigerian Inst. Oil Palm Res. **6**: 145–189 (1982). —Dransfield *et al.*, Genera Palmarum, ed.2: 155 (2008).

Single-stemmed or clustered, acaulescent to erect, massive, monoecious palms, dying after flowering. Stem of massive construction with short internodes, usually covered with rotting leaf bases and fibres, often with upward-pointing roots. Leaf reduplicate, pinnate, massive (some of the largest leaves in the plant kingdom); leaf base briefly sheathing, some species (not in the Flora area) producing broad black fibres clothing the stem; petiole short to long; rachis channelled adaxially in proximal region, 2 lateral grooves accommodating the leaflets in bud; leaflets numerous, crowded, one-fold, arranged in 1 to several planes; in most species leaflets spiny along margins and/or midvein, usually with abundant wax on lower surface. Inflorescences produced ± simultaneously in axils of the most distal leaves or bracts, erect or pendulous; axis of inflorescence bearing a prophyll and several empty tubular bracts, followed by subdistichous or faintly 4-ranked bracts each subtending a first order branch; first order branches each with a basal prophyll and several empty tubular bracts, followed by subdistichous or faintly 4-ranked bracts each subtending a rachilla; rachillae with female flowers in the proximal $^1/_4$–$^3/_4$, male flowers in furthest portion, each flower enclosed within a 2-keeled prophyll, female flower with a second tubular bracteole. Male flower with tubular, scarcely 3-lobed calyx; corolla tubular below, with 3 free lobes above; stamens 6–20 or more, briefly epipetalous and partly joined by their fleshy filaments; pistillode sometimes present, minute.

Female flower with tubular 3-lobed calyx; corolla tubular below, with 3 free petals above; staminodal ring present, epipetalous, with sterile flattened, sagittate anthers; pistil with 3 free or united terminal stigmas; ovary covered with vertical rows of reflexed fimbriate scales; locules 3, each with a single bitegmic, anatropous ovule, normally only 1 ovule developing to maturity. Fruit tipped with stigmatic remains and covered with enlarged reflexed scales arranged in vertical rows; mesocarp oily. Seed with a moderately thick dry testa and sparsely ruminate endosperm. Seedling leaf pinnate.

About 20 species, all except two confined to the wetter regions of Africa, one in Madagascar, and one in eastern South and Central America. Some species are cultivated outside their natural range.

All species are used extensively for construction of houses and furniture, thatch, food (sago can be obtained from the pith of the trunk) and oil (from fruit) and some species are tapped for wine.

Inflorescence pendulous, hanging down between leaves . . . . **1.** *farinifera*
– Inflorescence crowded into a tall, erect compound terminal inflorescence, overtopping leaves . . . . **2.** *australis*

1. **Raphia farinifera** (Gaertn.) Hylander in Lustgarden **31**: 91 (1952). —Dransfield in F.T.E.A., Palmae: 38 (1986). —Tuley in Palms Africa: 66 (1995). —White *et al.*, Evergr. For. Fl. Mal.: 107 (2001). —M. Coates Palgrave, Trees Sthn. Africa, ed.3: 101 (2002). Type: Illustration t.120 in Gaertner, De Fructibus 2 (1791). FIGURE 13.2.**9**.

   Sagus farinifera Gaertn., De Fruct. **2**: 186, t.120 (1791).
   Raphia ruffia (Jacq.) Mart., Hist. Nat. Palm. **3**: 217 (1838). —Warburg in Engler, Pflanzenw. Ost-Afr. **C**: 131 (1895). —Wright in F.T.A. **8**: 104 (1901). Type: Illustration t.4/2 in Jacquin, Fragm. Bot. **7** (1801).
   Raphia vinifera sensu Kirk in J. Linn. Soc. **9**: 234 (1866), non P. Beauv.
   Raphia kirkii Becc. in Webbia **3**: 58 (1910). Type: Zanzibar, n.d. *Kirk* (K lectotype).
   Raphia kirkii var. *grandis* Becc. in Webbia **3**: 64 (1910). Type: Malawi, 1896, *Johnston* (K lectotype).

Massive, clustering (rarely solitary) palm to 25 m or more; trunk to 60 cm or more in diameter, lower part with pronounced leaf scars, rotted leaf sheaths and upward-pointing adventitious roots, upper part covered with leaf bases. Leaves erect, slightly spreading, giving most crowns a characteristic 'shuttlecock' appearance, very large, to 20 m long; leaf sheathing at base, with ragged ligular edge; petiole rounded, to 1.5 m long and 20 cm in diameter, decreasing gradually to c.12 cm at insertion of lowermost leaflets; rachis orange-brown or almost crimson when young, with 2 lateral grooves near base, accommodating leaflets in bud. Leaflets up to c.150 on each side, to 100 × 8 cm, mostly inserted in two planes, the whole leaf appearing plumose, stiff, hardly drooping; margins and main veins with distally pointing spines to 3 mm long; lamina white-waxy below; main veins usually reddish. Inflorescences produced ± simultaneously from axils of reduced leaves at stem apex, pendulous, massive, to 300 × 35 cm wide. Primary inflorescence-bracts to 30 × 20 cm, tubular, partially enclosing first and second order branches; primary branching system to 30 × 2.5 cm wide at base, with a basal 2-keeled prophyll to 18 cm long; 1 or few empty tubular bracts to 6 cm long, further bracts c.5 mm apart, closely sheathing, ± 4-ranked, each subtending a rachilla, all rachillae lying close together, almost congested at anthesis, to 12 cm in lower part of inflorescence, shorter in furthest part, exceptionally to 15 × 1.5 cm. Male flower to 12 × 2 mm, enclosed in a 2-keeled prophyll to 6 mm long; calyx tubular, scarcely 3-lobed, to 4 mm long; corolla with basal tube to 2 mm long and 3 lobes 10 × 1.5 mm, acute, slightly thickened in uppermost 3 mm; stamens 6, epipetalous at mouth of corolla-tube, filaments 2 × 0.5 mm wide, fleshy and weakly joined, anthers c.3.5 × 0.5 mm; pistillode not seen. Female flower enclosed in a 2-keeled prophyll to 8 mm long, a second bracteole to 3 mm long; calyx tubular, truncate, usually splitting, to 8 × 4 mm; corolla

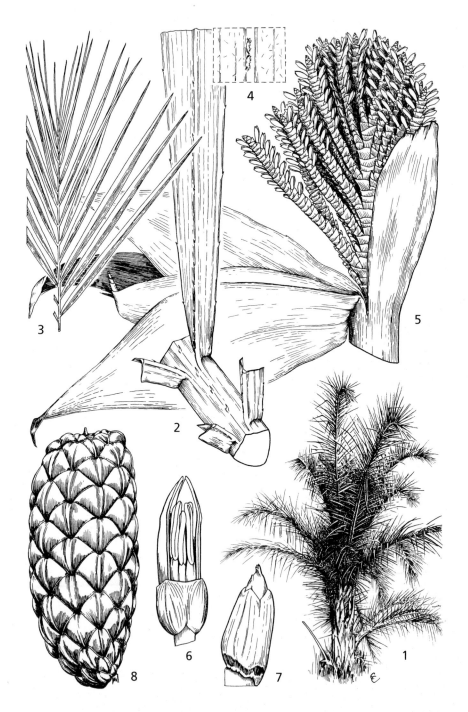

Fig. 13.2.**9**. RAPHIA FARINIFERA. 1, habit (× 2); 2, section of 1/200 leaf-rachis (× $^2/_3$); 3, leaf tip (× $^1/_6$); 4, detail of leaflet scale (× $1^1/_2$); 5, first order branch of inflorescence with rachillae (× $^1/_2$); 6, male flower, one petal removed (× 4); 7, female flower (× 4); 8, fruit (× $^2/_3$). 1 from photo, 2–4 from *Greenway* 5390, 5,8 from *Polhill & Paulo* 985, 6,7 from *McArthy & Eyles* 6678. Drawn by Eleanor Catherine. From Flora of Tropical East Africa.

much shorter than calyx, tubular in lower 2 mm, with poorly developed apiculate lobes to 3 × 2 mm; staminodes 6, inconspicuous; ovary to 5 × 3 mm, covered in vertical rows of reflexed orange-brown fimbriate scales. Fruit very variable, ovoid, pyriform or ellipsoidal, sometimes developing without seed (in which case very narrowly ellipsoidal), 7.5–10 × 4–5.5 cm, covered in 12–13 rows of convex reflexed scales, tipped by a short beak to 5 mm high; scales usually with pronounced mid-groove and deep dimple at base, which fits tip of next scale, largest scales c.15 × 18 mm wide, bright orange-brown; mesocarp to c.5 mm thick when fresh, oily, drying very hard. Seed usually 1, 4–6 × 3–4 cm wide, shape conforming to external shape of fruit, pointed at base.

Zambia. W: Mufulira Dist., 15.xi.1968, *Fanshawe* 10426 (K, NDO). Also sight records from N: (Luapula), B: (Mongu and Kalabo districts), C: (Serenje Dist., Kundalila Falls), E: (Kachalola) and S: (Chisekese) (M. Bingham, pers. comm.). **Zimbabwe**. N: Mt Darwin Dist., Great Dyke, Umvukwe hills, foot of Mvurwe Mt, 1370 m, 20.xii.1952, *Wild* 3938 (K, SRGH). E: Mutare Dist., Burma Valley, c.2 km upstream of Burma homestead on banks of Nyamataha R., 15.viii.1962, *Chase* 7791 (K, SRGH). **Malawi.** N: Rumphi Dist., Chelinda R. near Rumphi, 1100 m, 3.xii.1975, *Pawek* 10387 (K, MAL, MO, SRGH, UC). S: Phalombe Dist, Fort Lister Gap, by Nkombeni R., 1100 m, fr. 8.vi.2008, *Timberlake, Alves & de Sousa* 5362 (K, LMA, MAL). **Mozambique**. N: Marrupa Dist., Rio Messalo, 23 km along road Marrupa–Nungo, 1 km upstream from bridge, 7.viii.1981, *Jansen, de Koning & Wilde* 120 (K, LMA, LMU, WAG). Z: Milange Dist., Sabelua, river margin, 16°28'05"S 35°58'59"E, fr. 22.vi.2000, *Boane* 347 (LMU). T: Angónia Dist., Ulónguè, Rio Livelange, 17.xii.1980, *Macuácua* 1471 (LMA). MS: Manica Dist., between Chimoio and Garuso, 23.vi.1961, *Leach & Wild* 11128 (K, SRGH).

Throughout southern tropical Africa and Madagascar, where it is probably introduced; frequently receives a rudimentary cultivation. In gallery forest and freshwater swamp-forest; 400–1400 m.

Conservation notes: Locally common in the Flora area, but habitat under threat; Near Threatened. A protected species in Zimbabwe, where it is considered nationally to be Vulnerable B1B2C2 (Sabonet Red Data List, Golding 2002).

Variations in fruit size and shape were used by Beccari to separate *R. kirkii* from *R. ruffia* and varieties within *R. kirkii*. However, variation in these characters occurs within populations in the wild. In contrast, *R. farinifera* in Madagascar, where it is thought to be introduced, is remarkably invariable.

White *et al.* (2001) suggest that the distribution of this species is linked to that of the Palm-nut Vulture.

2. **Raphia australis** Oberm. & Strey in Bothalia **10**: 29 (1969). —Tuley in Palms Africa: 65 (1995). —M. Coates Palgrave, Trees Sthn. Africa, ed.3: 100 (2002). Type: South Africa, KwaZulu-Natal, Ingwavuma Dist., Kosi Bay, xi.1967, *Strey* 7785 (PRE holotype, K).

Massive, single-stemmed palm to 16 m or more; stem to 50 cm dbh, covered with closely adpressed old leaf bases, lower part becoming bare with leaf scars, remains of rotted leaf sheaths and adventitious roots. Leaves erect at first, then spreading, very large, to 10 m long or more; leaf sheathing at base, with ragged ligular edge; petiole rounded, to 1 m long and c.15 cm in diameter, decreasing gradually to c.10 cm at insertion of lowermost leaflets; rachis with 2 lateral grooves near base, accommodating leaflets in bud. Leaflets linear, up to 120 or more on each side, to 45–120 × 3–7.2 cm, mostly inserted in two planes, the whole leaf appearing plumose, stiff, hardly drooping; margins and upper surface of midrib with distally pointing spines to 7 mm long, leaflet tips bifid to 25 cm; lamina white-waxy below; main veins usually reddish. Inflorescences produced ± simultaneously from axils of reduced leaves and bracts at stem apex, producing a massive erect suprafoliar compound inflorescence to 3 m tall; reduced leaves subtending lowermost inflorescences to 80 cm. First order branches of inflorescence up to c.1.5

m long, branched to 2 orders, rachillae erect at first, becoming pendulous in fruit, in 4 ranks, to 30 × 1 cm in basal inflorescences, shorter and more slender towards tips, with female flowers in lower half and male flowers above; rachilla bracts c.8 mm long, mouths c.15 mm across, striate. Male flowers to 15 × 2.5 mm, enclosed by 2-keeled prophyll to 3 mm long; calyx tubular, scarcely 3-lobed, to 2 mm long; corolla with basal tube to 1 mm long and 3 lobes, 13 × 1.5 mm, acute, slightly thickened in uppermost 2 mm; stamens 6, epipetalous at mouth of corolla-tube, free filaments 4 × 0.5 mm, fleshy and weakly joined, anthers c.8 × 0.5 mm; pistillode not seen. Female flowers enclosed in 2-keeled prophyll to 8 mm long and second bracteole to 3 mm; calyx tubular, truncate, usually splitting, to 8 × 4 mm; corolla shorter than calyx, tubular in lower 2 mm, with triangular apiculate lobes to 3.5 × 1.5 mm; staminodes 6, inconspicuous, united in a ring joined with corolla; ovary to 4 × 2 mm, covered in vertical rows of reflexed fimbriate scales. Fruit ellipsoid, 5.5–9 × 3–5 cm, including a short beak 3.5–5 mm long, pericarp covered in 12 vertical rows of dark chestnut-brown scales with shallow central groove and paler fimbriate margins; mesocarp to c.5 mm thick, oily when fresh, drying very hard. Seed usually 1, 4–6 × 3–4 cm, its shape conforming to external shape of fruit, pointed at base.

Mozambique. M: Marrucuene Dist., 35 km N of Maputo, banks of Rio Boboli, 6.ii.1968, *Gomes e Sousa & Balsinhas* 5036 (K, LMA).

Confined to NE South Africa and S Mozambique. In gallery forest and freshwater swamp-forest; 0–50 m.

Conservation notes: Known from only one locality in the Flora area; globally considered Data Deficient (IUCN Red Data List 2009) but probably Vulnerable.

Instantly recognizable when in flower by the huge terminal erect compound inflorescence. The only other species of *Raphia* with a suprafoliar inflorescence is *R. regalis* Becc., an acaulescent species from the humid forests of Cameroon and Gabon which has the largest leaf in the plant kingdom (25 m or more long).

3. **CALAMUS** L.

Calamus L., Sp. Pl.: 325 (1753). —Dransfield *et al.*, Genera Palmarum, ed.2: 191 (2008).

Solitary or clustering, acaulescent to high-climbing dioecious palms, not dying after flowering. Stems very slender to robust, in Africa mostly 1–3 cm in diameter, branching sympodially at base. Leaf sheaths tightly enclosing stem, variously armed with spines or unarmed; often continued into an ocrea of varying length. Mature leaves of two kinds, either terminating in a long barbed whip (cirrus) or without (African and some Asiatic species), species without a cirrus normally with a similar barbed whip (flagellum) joined to leaf sheath at its base, equivalent to a modified sterile inflorescence; petiole prominent or absent, variously armed with spines and hooks; rachis usually armed with reflexed hooks. Leaflets narrow to broad or rhomboid, single-fold, arranged on either side of rachis, or variously clustered, fanned or paired, variously hairy, scaly or spiny. Inflorescence axillary, base of peduncle joined to internode and sheath of the following leaf, so appearing in a non-axillary position, branching 2–4 orders, with or without a long terminal flagellum; bracts variously armed, tubular, tightly sheathing, rarely splitting, sometimes with an expanded limb, never caducous; prophyll usually 2-keeled and empty; other bracts on main axis subtending the partial inflorescence; partial inflorescences with bracts subtending rachillae, rachillae usually with approximate tubular bracts, each (except for basal) subtending a flower or flower group; male inflorescence with flowers solitary, borne together with a bracteole ('involucre'); female inflorescence with flowers borne in pairs, a sterile male (acolyte) and a fertile female and 2 bracteoles ('involucrophore' and 'involucre'). Male flowers symmetrical; calyx tubular, 3-lobed; corolla 3-lobed, divisions almost reaching base; stamens 6, epipetalous, with free filaments; pistillode minute or absent. Female flowers with calyx and corolla as in male; staminodes 6; ovary covered in vertical rows of reflexed scales, tipped with 3 stigmas; locules 3, incomplete, each with single ovule, normally only 1 developing. Sterile female flower as the fertile, but anthers empty. Fruit variously shaped, tipped with stigmatic remains, with a persistent calyx and corolla at base, covered in vertical rows of reflexed scales.

A genus of about 374 species distributed from the Himalayas, S China and Taiwan through SE Asia, the Malay Archipelago to Australia, Fiji and Vanuatu; a single very variable species in Africa.

The genus is of immense significance as the most important source of rattan canes used in the furniture and handicraft industries, but also with a wide range of local uses. The most important economic species are all Asiatic.

Calamus deerratus G. Mann & H. Wendl. in Trans. Linn. Soc. **24**: 429 (1864). — Wright in F.T.A. **8**: 108 (1901). —Beccari in Ann. Roy. Bot. Gard. Calc. **11**: t.19 (1908). —Russell in F.W.T.A., ed.2 **3**: 166 (1968). —Dransfield in F.T.E.A., Palmae: 43 (1986). —Tuley in Palms Afr.: 50 (1995). —Burkill in Useful Pl. W Trop. Afr. **4**: 347 (1997). —Sunderland, Field Guide Rattans Afr.: 62 (2007). Types: Sierra Leone, Bagroo R., iv.1861, *Mann* 895 (K syntype, FI); Cameroon R., i.1863, *Mann* 2147 (K syntype). FIGURE 13.2.**10**.

    Calamus laurentii De Wild. in Ann. Mus. Congo, Bot. **5**: 97 (1904). —Durand & Durand in Fl. Cong. **1**: 584 (1909). —Beccari in Ann. Roy. Bot. Gard. Calc. **11**, app.: t.2 (1914). Type: Congo, Eala, n.d., *Laurent* 126 (BR holotype, FI).

Clustering, high-climbing, spiny palm to 20 m or more. Stems branching sympodially at base, 7–30 mm in diameter; internodes 8–20 cm long. Leaf sheaths very varied in armature from ± unarmed to densely spiny, with marked knee below petiole; spines black, flattened, up to 3 cm long, occasionally grouped, sometimes clusters of upward pointing spines around leaf sheath mouth; ocrea present, to 12 cm long, usually less, usually conspicuous, dry, papery, tongue-shaped, then splitting and becoming bilobed, armed with paler spines, rarely unarmed. Leaves lacking a terminal whip, to 175 cm long, usually less; petiole to 20 cm, rounded below, channelled above, c.5 mm broad, variously armed with large black 3 cm spines and small recurved black hooks; rachis triangular in section above; leaflets ± concolorous, numerous, up to 30 on each side of the rachis, usually grouped in 3s to 5s above, up to 35 × 2 cm, widest at ± ¹/₃ of length, tapering to a long tip, margins and main vein bristly throughout, somewhat pleated with up to 9 prominent secondary nerves and prominent sinuous lateral veins. Flagellum to 2 m long by 4 mm wide at the base, decreasing very gradually above, armed with small recurved hooks. Male and female inflorescences to 2 m long, similar, with 1–4 partial inflorescences and a long terminal sterile flagellum; axis and bracts armed with reflexed, solitary or grouped black spines; bracts tightly sheathing, up to 70 cm long, with an expanded, somewhat papery limb c.5 cm long; partial inflorescences to 40 cm long, up to 15 rachillae on each side, subtended by bracts c.2 cm long (1 cm exposed), mouths c.7 mm wide, with short triangular limb to 4 mm; rachillae to 7 cm long, usually arcuate, arranged distichously; bracts distichous, dull brown, ± ciliate-hairy around mouth. Male flowers solitary, distichous, with a minute involucre to 1 mm long; calyx 4 mm long, tubular for ¹/₄ its length, with 3 short, triangular, striate lobes; corolla-lobes to 7 × 2 mm, joined at base for c.1 mm, widely diverging at anthesis; stamens to 4 mm long, minutely epipetalous, filaments to 3 mm long, anthers c.3 mm long, medifixed; pollen yellow. Sterile male flower very similar to fertile ones but slightly shorter and narrower. Female flowers with calyx tubular at first, then splitting as ovary increases in size, lobes c.3 mm long; corolla-lobes c.5 × 2 mm, with 6 minutely epipetalous flattened staminodes; ovary c.5 × 2.5 mm, tipped by 3 stigmas c.1 mm long, markedly recurved at anthesis. Fruit to 1.5 × 1 cm, with a short beak to 2 mm tipped by stylar remains, (15)17–18(20) vertical rows of dull, pale brown scales, edged with darker brown. Seed somewhat flattened laterally, c.9 × 8 × 5 mm.

Zambia. N: Chiengi Dist., Luau R., Chicuga, 1200 m, 16.xii.1989, *Vongkaluang* s.n. (K). Widespread throughout humid Equatorial Africa, usually in swampy places; 1200 m.

Conservation notes: Known from only a single collection in the Flora area, where it is probably Vulnerable or Endangered.

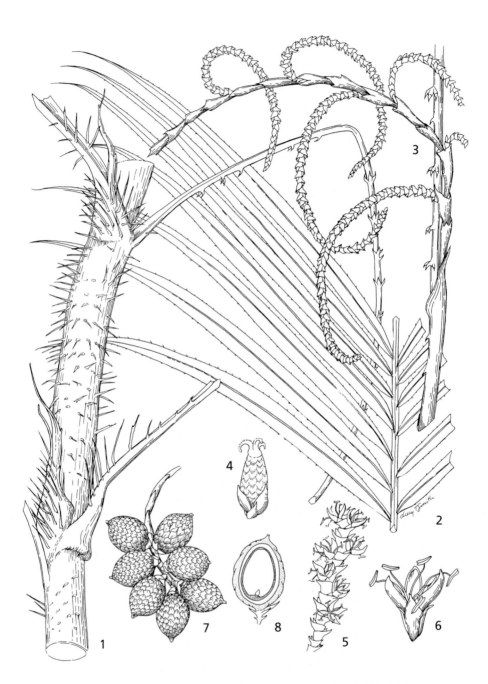

Fig. 13.2.**10**. CALAMUS DEERATUS. 1, stem (× ¹/₂); 2, portion of leaflets (× ¹/₃); 3, inflorescence (× ²/₃); 4, male flower (× 2²/₃); 5, male flower on rachilla (× 1); 6, female flower (× 2²/₃); 7, fruit (× 1); 8, fruit, longitudinal section (× 1¹/₂). 1–2 from *Sunderland* 2262, 3–5 from *Sunderland* 1754, 6 from *Eggeling* 1626, 7–8 from *Deighton* 1847. Drawn by Lucy Smith. From Sunderland, Field Guide Rattans of Africa (2007).

There is an unconfirmed record for Zambia W: Mwinilunga Dist., swamp forest, in Fanshawe (Veget. Mwinilunga Dist., For. Res. Pamphl. 27, 1969).

Several taxa have been described, but current thinking is that they all belong to one variable species. Juvenile stems can appear to be different from adult stems nearby.

4. PHOENIX L.

Phoenix L., Sp. Pl.: 1188 (1753). —Barrow in Kew Bull. **53**: 513–575 (1998). — Dransfield *et al.*, Genera Palmarum, ed.2: 242 (2008).

Solitary or clustering, acaulescent, shrubby or erect dioecious palms. Trunk usually covered in persistent leaf bases when young, later smooth. Leaves induplicately pinnate; leaf sheath fibrous; leaflets numerous, single-fold, usually sharply pointed, the lower ones small, modified as sharp spines (acanthophylls), the upper ones held in 1 to several planes, surfaces frequently waxy. Inflorescences interfoliar, branching to one order only; peduncle short or elongating after anthesis, bearing a 2-keeled prophyll enclosing inflorescence in bud; peduncular bracts absent; rachillae often arranged spirally or in partial spirals, each subtended by a small bract. Flowers borne singly in a spiral along the rachilla, each subtended by an inconspicuous bract. Male flowers usually asymmetrical; calyx cup-shaped with 3 short triangular lobes; corolla briefly tubular with 3 lobes, ± valvate, much exceeding the calyx; stamens (3)6(9), epipetalous, with short filaments and elongate anthers; pistillode absent or with 3 minute carpel vestiges. Female flowers dissimilar, globular; calyx cup-shaped with 3 short triangular lobes; corolla with 3 free petals, imbricate, usually rounded; staminodes 6, minute, sometimes united in a short cup; carpels 3, free, somewhat elongate, with short recurved fleshy stigmas; ovule anatropous, basally attached, usually only one carpel developing into a fruit. Fruit with smooth epicarp, sometimes waxy; mesocarp fleshy; endocarp consisting of a thin membrane. Seed 1, basally attached, deeply grooved longitudinally.

Genus of 13 species in the Old World tropics and subtropics, extending from the Canary Islands and most of Africa to Crete, Turkey, Madagascar, through Arabia and the Indian subcontinent to SE Asia, extending south to 2°N in the Malay Peninsula and Sumatra; one species in southern Africa.

Phoenix dactylifera L., the common cultivated Date Palm, can easily be distinguished from *P. reclinata* by the much more massive trunk, glaucous leaves, and the leaflets all distinctly distally pointing. Superficially rather similar is the Canary Islands Date Palm, *P. canariensis* Chabaud, sometimes cultivated in upland areas. This however is even more massive, with a very large trunk 75–100 cm or more in diameter, and it has dark green, not glaucous leaflets, abscissing to leave very broad rather low leaf scars rather than the narrower, taller scars of *P. dactylifera*. When seen together the differences are striking.

The Date Palm is one of the oldest domesticated plants and is widely distributed across the margins and oases of the Sahara and into the Arabian Peninsula. It is now introduced to California and other places with a suitable climate. Other species are widely used, particularly as a source of fibre for basket weaving. The fruit of all species is edible, but the flesh is often too thin to be of any importance as a food crop.

Phoenix reclinata Jacq., Fragm. Bot.: 27, t.24 (1801). —Wright in F.T.A. **8**: 103 (1901). —White, For. Fl. N. Rhod.: 12 (1962). —Russell in F.W.T.A., ed.2 **3**: 169 (1968). —Tuley in Palms Africa: 18 (1995). —White *et al.*, Evergr. For. Fl. Mal.: 107 (2001). —M. Coates Palgrave, Trees Sthn. Africa, ed.3: 97 (2002). Type: Illustration t.24 in Jacquin, Fragm. Bot.: 27 (1801). FIGURE 13.2.**11**.

Clustered, rarely solitary palm, often forming dense thickets with trunks to 10 m or more tall, c.15 cm dbh, occasionally flowering while still acaulescent. Trunk dull brown, free of leaf-sheaths below, with persistent leaf-sheaths above, uppermost 1–2 m irregularly marked with oblique leaf scars, with persistent leaf-bases; injured stems exuding a clear yellowish gum. Crown of 25 green leaves or more, dead leaves frequently rather long-persistent. Leaves to 2.5 m long, bright shiny green, not glaucous, arcuate; true petiole c.15 cm, with coarse red-brown sacking-like sheath; apparent petiole c.50 × 2.5 cm, armed with fanned irregularly arranged acanthophylls to 6 × 0.4 cm; leaflets ± 120 on each side of the rachis, arranged very stiffly and regularly above, grouped in 2–4s below, fanned, to 25 × 2 cm, pointed, occasionally sharp, old leaflets splitting along the main vein, young leaflets with white indumentum on lower surface. Inflorescence with prophyll 24–50 × 7–10 cm, often persistent, frequently splitting longitudinally into 2 halves, bright orange-brown at anthesis, fading to dull grey-brown, covered with flocculent grey-brown indumentum when young, soon falling; peduncle not greatly elongating, sometimes scarcely emerging from bract, 10–30 × 1.5 cm, greatly compressed; rachillae 40–70, arranged in groups and partial spirals, to 15 cm long, usually much less, 2.5 mm wide near base tapering to 1.5 mm above. Male flowers creamy white, rapidly turning brown, musty scented; calyx c.1 mm high; petals 6–7 mm long, acute, fleshy, somewhat dentate towards tip, tip slightly reflexed at anthesis; stamens 6, slightly shorter than petals, pale brown. Female inflorescence emerging from bract, often greatly elongating after anthesis, fruiting rachillae pendulous; rachillae 40–60, 15–40 cm long, with up to 40 flowers, singly or in small groups. Female flowers greenish, rounded, c.2 mm wide; calyx c.1.5 mm high; petals rounded c.2 × 2 mm; carpels 3, c.2 mm high, just emerging from the tightly imbricating petals, stigmas reflexed. Fruit 1.3–1.7 × 0.9–1.3 cm, usually developed from only 1 carpel, rarely all 3 developing; calyx to 2 mm high in fruit, petals to 5 × 8 mm wide, varying from pale yellow to orange or dull red; epicarp smooth; mesocarp 1–2 mm thick, dry or moist and sweet. Seed 1–1.2 × 0.6–0.8 cm, deeply grooved along one side.

Caprivi. Kakumba Is., 17.i.1959, *Killick & Leistner* 3418 (SRGH). **Botswana**. N: Okavango, island in Zibadianja lagoon, 20.x.1972, *Biegel, Pope & Gibbs Russell* 4018 (K, SRGH). **Zambia**. B: Senanga Dist., Kaunga, near Mashi R., 18.x.1962, *Mubita* in *SRGH* 138,918 (K, SRGH). N: Mbala Dist., Mulungu R. swamp, 1500 m, 6.x.1956, *Richards* 6387 (K). W: Solwezi Dist., Meheba R., 23.vii.1930, *Milne-Redhead* 758 (K). C: Luangwa Dist., Katondwe, 30°15'E 15°15'S, 400 m, 13.iv.1988, *Phiri* 2418 (K, UZL). S: Namwala Dist., banks of Kafue R., near Baunza, 25.ix.1963, *van Rensberg* KBS 2484 (K, SRGH). **Zimbabwe**. N: Lomagundi Dist., bottom of Mutorashanga Pass, 29.iii.1969, *Pope* 42 (K, SRGH). W: Hwange Dist., Victoria Falls, near Palm Grove, 19.xi.1949, *Wild* 3109 (K, SRGH). C: Wedza Dist., Wedza Mt, 22.v.1968. *Mavi* 764 (SRGH). E: Mutare Dist., N Penhalonga, Odzani R. valley, 18°46.2'S 32°41.7'E, 1600 m, 19.xii.1994, *Wilkin* 724 (K, SRGH). S: Masvingo Dist., Mushandike Nat. Park, below dam wall, TN514 704, 29.viii.1983, *Mahlangu* 783 (SRGH). **Malawi**. S: Blantyre Dist, Bangwe hill, 4 km E of Liombe, 1260 m, 25.xi.1977, *Brummitt, Seyani & Banda* 15156 (K, MAL). **Mozambique**. N: Palma Dist., Palma–Pundanhar road, 4.5 km W of junction with Nhica do Rovuma road (sight record, D. Goyder *et al.*, 2009). Z: Mopeia Dist., between Conho and Mopeia, 29.xii.1969, *Amico & Bavazzano* s.n. (FIR, K). T: Angónia Dist., Ulónguè, fl. 26.xi.1980, *Macuácua* 1325 (LMA). MS: Gondola Dist., Inchope, on road to Rio Revué, fr. 3.xi.1953, *Gomes Pedro* 4548 (LMA). GI: Massinga Dist., Pomene, around hotel, fl. 21.ix.1980, *Jansen, de Koning & Zunguze* 7448 (LMA, LMU). M: Matutuíne Dist., Bela Vista, Tinonganine, 11.xii.1961 *Lemos & Balsinhas* 291 (LMA).

Found throughout the moister parts of sub-Saharan tropical Africa, almost reaching the Cape of Good Hope, Somalia, Saudi Arabia and Yemen. Particularly beautiful stands occur at Victoria Falls. Widely planted in Zimbabwe as a garden plant. At lower altitudes tending to grow along watercourses, but in higher rainfall areas and mountains growing also on open rocky hillsides, cliffs and even in forest, but confined to areas with a sparse canopy; 0–1600 m.

Conservation notes: Very widespread, with some local over-exploitation of its leaves for weaving; not threatened.

Fig. 13.2.**11**. PHOENIX RECLINATA. 1, habit (× $^1/_{150}$); 2, leaf base with acanthophylls (× $^2/_3$); 3, basal part of rachis (× $^2/_3$); 4, middle part of rachis (× $^2/_3$); 5, leaf tip (× $^2/_3$); 6, prophyll (× $^2/_3$); 7, fragment of male inflorescence (× $^2/_3$); 8, male flower with one petal removed (× 4); 9, part of female inflorescence (× $^2/_3$); 10, female flower (× 4); 11, fruit (× 2); 12, seed (× 2). 1 from photo of *Wild* 3109, 2,9,10 from *Dransfield* 4828, 7–8 from *Dransfield* 4829, 11–12 from *Dransfield* 4832. Drawn by Eleanor Catherine. From Flora of Tropical East Africa.

A very polymorphic species in its vegetative characters; some montane forms even resemble *P. rupicola* T. Anders. of India, apart from being clustered. There is a whole range of intermediates between the montane and lowland forms, but little variation in details of flower and fruit.

Widely used throughout the region as a source of material for fine weaving. Leaflets of the unexpanded "sword leaf" are sun dried, split, and used in plaited weaving for mats and baskets. The petiole and rachis are used, split, in coarse weaving and in the production of cordage. The fruits are rarely eaten.

5. HYPHAENE Gaertn.

Hyphaene Gaertn., De Fruct. **1**: 28 (1788). —Beccari, Palme Borasseae: 18–49 (1924). —Furtado in Garcia de Orta **15**: 427–460 (1967). —Dransfield *et al.*, Genera Palmarum, ed.2: 314 (2008).

Small to robust, solitary or clustering dioecious palms, not dying after flowering. Trunk usually branching by repeated forking, apparently truly dichotomous, forking sometimes occurring underground giving rise to a clump of even-sized trunks, clothed when young with persistent leaf bases, then rotting or succumbing to fire leaving a clean trunk. Leaves borne spirally; leaf base partially sheathing, split to produce a central triangular cleft; petiole usually well-developed, semi-circular in cross-section, densely armed with reflexed or upward-pointing spines, often with stellate scales and wax; upper hastula conspicuous except in very young seedling leaves, frequently asymmetrical, markedly oblique, often partially obscured by dense hairs and scales in young leaves; blade often conspicuously costapalmate, divided to ± $^1/_3$ its radius into single-fold induplicate segments, filaments often conspicuous at sinuses; lamina-surfaces frequently waxy, dotted with black scales; longitudinal and transverse veinlets inconspicuous. Inflorescences similar, male frequently more slender and highly branched than female; axis frequently flattened towards base, with a basal, empty, 2-keeled tubular prophyll and 1–2 tubular peduncular bracts with short triangular limbs; rachis longer than peduncle; bracts frequently densely stellate-hairy; first-order branches usually markedly compressed, semi-circular in cross-section, elongate, bearing 1–13 rachillae at tip (usually 1 or 2 in female) in a fascicle, each subtended by a small triangular bract; rachillae with imbricate, spirally arranged, laterally connate bracts, each enclosing a floral pit, often densely filled with fluffy hairs. Male flowers in groups of 3 forming a cincinnus, embedded in hairs, one flower exposed at a time; calyx tubular at base, with 3 apical cucullate lobes; corolla stalk-like at base, with 3 apical triangular imbricate lobes, reflexed at anthesis; stamens 6, at base of corolla-lobes with short filaments and basifixed anthers; pistillode minute. Female flowers solitary in axil of each bract on a densely pubescent pedicel, much larger than male; sepals 3, free, imbricate, rounded; petals 3, free, imbricate, rounded, similar to sepals; staminodal ring membranous, with 6 teeth, each tipped with minute empty anthers; ovary globose, with 3 apical, triangular, ± sessile stigmas, 3-locular, usually only 1 ovule developing to anthesis, hence ovary asymmetrical, occasionally all carpels equally developed. Fruit extremely variable, borne on enlarged pedicel, the perianth-segments persistent but hardly enlarging; stigmatic remains basal; epicarp sometimes pitted, various shades of brown; mesocarp fibrous, often aromatic and edible; endocarp hard, stony; endosperm homogeneous, hollow; embryo apical. Germination remote-tubular; seedling leaf simple, plicate.

Genus of 8 species widespread in subtropical and tropical Africa, though avoiding the wettest and driest areas, the Arabian Peninsula, west coast of Indian subcontinent, and 1 species from Sri Lanka, though doubtfully native.

The genus *Hyphaene* has been much misunderstood and confused. Because of the nature of variation and the scrappy material available, too many names have been created based on single fruits, sometimes without reference to habit or locality. In the field, many workers have experienced difficulty in applying the names. More fieldwork elsewhere in Africa is necessary before monographic treatment of the genus can be undertaken. In particular *Hyphaene* in NE Africa appears very complex; some taxa in this area may intergrade with those of occurring further south. Determinations of herbarium material without ripe fruit or accompanying field notes on habit and photographs are unsatisfactory and unreliable, even though mature plants in the field are usually very characteristic. It is also difficult to name juvenile specimens in the field, although in well-grown juveniles, subtle differences in leaf shape and colour provide useful distinguishing characters. Wild populations are nearly always influenced by humans to some extent, by burning, tapping for wine and cutting for leaf fibre, adding to the difficulty of distinguishing species. There is also a possibility that hybridization between taxa may take place.

All species of *Hyphaene* are widely and intensively utilised for raw material for thatching, cordage and matting, and are tapped for palm wine. The fruits are sometimes eaten and are relished by animals such as elephant, which are probably the most significant natural dispersers.

1. Stem erect, usually solitary, almost always unbranched, often swollen; ripe fruit rounded, without humps, ridges or swellings, except at base near pedicel; epicarp highly polished, smooth, pitting scarcely visible . . . . **1.** *petersiana*
- Stem erect, almost always branched, or low shrubby palms; fruit variously compressed, swollen, humped, ridged or misshapen; epicarp pocked or matt, not highly polished 2
2. Tall tree palm, solitary stem usually dichotomising many times, or dichotomising below ground, showing even-sized curving stems; ripe fruit exceeding 6 cm long, usually laterally compressed on 2 faces, strongly pitted, often highly aromatic . . . . **2.** *compressa*
- Shrubby to small tree palm, rarely dichotomising more than twice, usually with decumbent stems, clustering at base; fruit rarely exceeding 6 cm long, obovoid to irregular, expanded distally with matt epicarp, not deeply pitted, scarcely aromatic . . . . **3.** *coriacea*

1. **Hyphaene petersiana** Mart., Hist. Nat. Palm., ed.2 **3**: 227 (1845). —Peters, Reise Mossamb., Bot. **2**: 508 (1864). —Tuley in Palms Africa: 29 (1995). —M. Coates Palgrave, Trees Sthn. Africa, ed.3: 99 (2002). Type: Mozambique, n.d., *Peters* s.n. (B holotype). FIGURE 13.2.**12**.

   Hyphaene benguelensis Welw. in Gaz. Med. Lisb. **16**: 451 (1862). —Furtado in Garcia de Orta **15**: 442 (1967). Type: Angola, Mossamedes Dist., Coroca R. near Porto Pinda, n.d., *Welwitsch* 6656 (LISU holotype, K).
   Hyphaene ventricosa J. Kirk in J. Linn. Soc. **9**: 235 (1866). —Wright in F.T.A. **8**: 122 (1901). —White, For. Fl. N. Rhod.: 12 (1962). —Fanshawe in Kirkia **6**: 105 (1967). Type: Zambia/Zimbabwe, Victoria Falls, not traced, possibly an illustration by T. Baines.
   Hyphaene goetzei Dammer in Bot. Jahrb. **28**: 354 (1900). Type: Tanzania, Ruaha R., i.1899, *Goetze* 413 (B† holotype, FI).
   Hyphaene aurantiaca Dammer in Bot. Jahrb. **30**: 267 (1901). —Wright in F.T.A. **8**: 122 (1901). Type: Tanzania, Lake Rukwa, 20.viii.1899, *Goetze* s.n. (B† holotype, FI).
   Hyphaene ventricosa subsp. *benguelensis* (Welw.) Becc., Palme Borass.: 44 (1924).
   Hyphaene ventricosa subsp. *ambolandensis* Becc., Palme Borass.: 44 (1924). Type: Namibia, Olukonda, n.d, *Schinz* (Z holotype, FI).
   Hyphaene sp. 1 of White, For. Fl. N. Rhod.: 12 (1962).
   Hyphaene benguelensis var. *ventricosa* (J. Kirk) Furtado in Garcia de Orta **15**: 446 (1967). —Drummond in Kirkia **10**: 232 (1975).
   Hyphaene obovata Furtado in Garcia de Orta **15**: 456 (1967). Type: Mozambique, Gorongosa Dist., Urema R., 4.v.1942, *Torre* 4046 (LISC holotype, COI, K, LMU).
   Hyphaene ovata Furtado in Garcia de Orta **15**: 457 (1967). Type: Zimbabwe, Matabeleland, no locality, t.6 in Pardy, Rhod. Agric. J. **52** (1955).

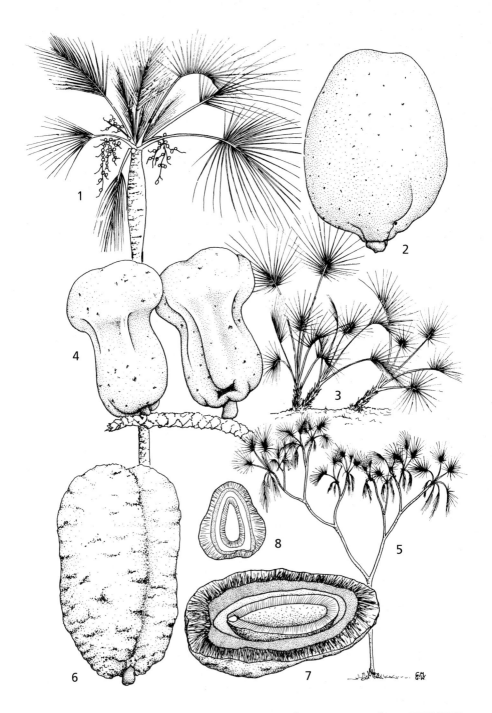

Fig. 13.2.**12**. HYPHAENE PETERSIANA. 1, habit (× $^1/_{80}$); 2, fruit (× $^2/_3$). HYPHAENE CORIACEA. 3, habit (× $^1/_{40}$); 4, fruits (× $^2/_3$). HYPHAENE COMPRESSA. 5, habit (× $^1/_{150}$); 6, fruit (× $^2/_3$); 7, fruit in longitudinal section (× $^2/_3$); 8, fruit in transverse section (× $^1/_3$). 1 from photo by Dyer, 2 from *Coombes* 1920, 3 from photo by Medley Woods (1890), 4 from *Dransfield* 4848, 5 from photo in Tanzania, 6–8 from *Dransfield* 4830. Drawn by Christine Grey-Wilson. From Flora of Tropical East Africa.

Solitary or rarely clustered (by dichotomy below ground) palm to 20 m tall, trunk almost always unbranched, to 35 cm dbh, sometimes with a swelling c.10 m above ground, then tapering to crown; leaves on young trees long-persistent, eventually rotting or burned to give a clean, grey, vertically cracked trunk marked with horizontal leaf scars, c.2.5 cm apart; crown with 20–25 green leaves. Leaf base broadly sheathing with a central triangular cleft to c.20 cm long; petiole 1–1.8 m long, c.10 cm wide at base, tapering to c.6 cm near insertion on lamina, channelled above near base, semicircular in cross-section below, armed with black forward-pointing spines along margins, with scattered black scales, buff hairs and wax; hastula oblique or symmetrical, to 5 mm high, blackish, usually armed with black spines and abundant buff hairs; lamina costapalmate, with costa curving to 75 cm long, lamina spread of 1.5–1.9 m, usually distinctly folded along costa, divided to ⅓–½ its radius into 35–40 segments c.5 cm wide, further divided for a short distance at tip along abaxial folds; inter-leaflet filaments prominent; whole lamina glaucous, dotted with small brown-black rounded scales, with buff scales along ribs. Male inflorescence arcuate, 1–1.7 m long; peduncle c.7 cm wide at base, with up to 12 partial inflorescences; lowermost bracts to 1.6 m long, densely waxy, scaly and hairy when young; partial inflorescences with up to 13 rachillae radiating in a semicircle, moderately stiff, sometimes sinuous; rachillae to 40 × 1.2 cm, with bracts c.7 × 2 mm, exposed at anthesis. Male flowers with narrow sepals, c.5 × 1 mm; corolla stalk c.3 mm high, lobes ovate, c.4 × 3 mm wide; filaments c.1 mm long, anthers yellow, c.1.5 mm long; pistillode minute. Female inflorescence arcuate, then pendulous, 1–1.25 m long, with up to 10 partial inflorescences, otherwise as the male; rachillae rarely more than 3 in each partial inflorescence, often 1 only, to 35 × 1.3 cm, pits c.8 × 5 mm at anthesis, filled with dull reddish-brown hairs. Female flowers with pedicel c.4 × 2 mm at anthesis, greatly increasing after fertilization to c.10 × 8 mm at fruit maturity, including the dense hairs; sepals ovate, c.3 × 3 mm; petals rather narrow, c.3 × 2 mm; ovary c.4 mm wide, green, stigma with nectar drop at anthesis. Ripe fruit variable, 5–8 × 5–7 × 5–6 cm, but always ± rounded, obovoid or ovoid, never regularly compressed, cottage-loaf shaped, swollen or warted except for low swellings at the base by pedicel, rarely with a very slight vertical ridge, sometimes with slight compression marks where very close-packed; epicarp very smooth, highly polished, with minute inconspicuous pitting, rich red-brown to chestnut, rather uniformly coloured, rarely with colour flecking, fragile at maturity, easily separating from mesocarp and its fibres; mesocarp aromatic, 5–10 mm thick; endocarp generally conforming to fruit shape, 5–7 mm thick. Seed top-shaped, broad end basal, up to 3.5 × 3.5 cm; endosperm c.6 mm thick.

Caprivi Strip. sight records. **Botswana**. N: Toromoja, Boteti R., st. 26.iv.1975, *Ngoni* 453 (K, SRGH). SE: Makgdaikgadi pans, margin of Sua Pan, Peterhouse Exped., 15.i.1974, *Kockott* 359 (SRGH). **Zambia**. B: Kalabo Dist., Kaunga, near Mashi R., 6.viii.1962, *Mubita & Reynolds* B153 (SRGH). N: Mpika Dist., Mfuwe, 11.v.1965, *Mitchell* 2911 (SRGH). C: Luangwa Dist., road to Luangwa R., c.25 km from Great East Road junction, by Mozambique border, fl.& fr. 21.x.1990, *Boyce & Cooper* DPP3 (K). Also sight records from E: and S: (M. Bingham, pers. comm.). **Zimbabwe**. N: Darwin Dist., Muzarabani, Musingwa R., 4.v.1972, *Mavi* 1393 (SRGH). W: Hwange Dist., Victoria Falls, by Zambezi R., fl. 18.xi.1949, *Wild* 3099 (K, SRGH). C: Makoni Dist., Mayo ICA, Koedoesrand Farm, fl. 15.xii.1968, *Avery* 1 (K, SRGH). E: Chimanimani Dist., Hot Springs resort (sight record, B. Wursten, 2006). S: Chiredzi Dist., Gonarezhou Nat. Park, near Save–Lundi junction, 250 m, st. 31.v.1971, *Ngoni* 141 (K, SRGH). **Malawi**. N: Karonga Dist., 24 km S of Karonga at lagoon, fr. 6.i.1978, *Pawek* 13547 (K, MAL, MO, SRGH, UC). S: Mangochi Dist., 29 km SW of Mangochi, edge of Lake Malombe, st. 24.ii.1979, *Brummitt & Patel* 15447 (K). **Mozambique**. N: Negamano Dist., 7 km towards Ngapa from Unity Bridge, Rio Rovuma, 150 m, 27.xi.2009, *Luke & Luke* 13934 (EAH, LMA). MS: Gorongosa Dist., Gorongosa Nat. Park, Chitengo camp, 30 m, 11.x.2007, *Ballings & Wursten* 907 (LMU).

Occurring from N Tanzania southwards to South Africa (former N Transvaal); also in Congo to the W coast in Angola and Namibia. Usually an inland taxon, often on alkaline soils with relatively high water-tables; 0–1300 m.

Conservation notes: Widespread and common in the Flora area; not threatened.

This species is sometimes confused with *Borassus aethiopum*, the two taxa being easily distinguished by the well-defined petiole-spines in *Hyphaene petersiana* vs. erose spines in *Borassus*, and the elongate pointed segments in *Hyphaene* vs. the stiffer, broader, more rounded segments in *Borassus*. When fertile, they are very easily distinguished by the inflorescence characters described above. With dried leaf fragments, all species of *Hyphaene* are immediately recognized to the genus by the presence of dark rounded scales scattered over the lamina, with very indistinct lateral veins; in *Borassus* there are no such scales, and lateral veins are distinct.

2. **Hyphaene compressa** H. Wendl. in Bot. Zeit. **36**: 116 (1878). —Wright in F.T.A. **8**: 123 (1901). —Tuley in Palms Africa: 26 (1995). Type: Vegetable ivory sample of unknown origin, not traced; Tanzania, Kilwa, *Zimmermann* 1ab (FI neotype), chosen by Beccari (1924).

    Hyphaene crinita sensu Wright in F.T.A. **8**: 121 (1901), non Gaertn.

    Hyphaene multiformis Becc. 'synspecies', Palme Borass.: 32 (1924).

    Hyphaene multiformis subsp. *rovumensis* Becc., Palme Borass.: 34 (1924). Type: Tanzania, Rovuma, n.d., *Busse* 1167 (B holotype, FI).

    Hypahene thebaica sensu Gomes e Sousa, Dendrol. Moç. **1**: 179 (1966), non Mart.

    Hyphaene incoje Furtado in Garcia de Orta **15**: 453 (1967). Type: Mozambique, between Quionga and Macanga beach, 17.iv.1964, *Torre & Paiva* 12128 (LISC holotype, K, SRGH).

    Hyphaene kilvaensis (Becc.) Furtado in Garcia de Orta **15**: 453 (1967). Type: Mozambique, Pemba, on coast, ix.1936, *Torre* s.n. (COI).

    Hyphaene megacarpa Furtado in Garcia de Orta **15**: 454 (1967). Type: Mozambique, Niaiba, between Imala and Muecate, 17.i.1964, *Torre & Paiva* 10035 (LISC holotype, K).

    Hyphaene semiplaena (Becc.) Furtado var. *gibbosa* (Becc.) Furtado in Garcia de Orta **15**: 458 (1967). Type: Mozambique, no locality, unknown collector (COI).

Massive tree palm, eventually reaching up to 20 m. Trunk solitary and erect, or apparently forking below ground giving a group of 2–4 curving trunks of ± same size, sometimes with clusters of young shoots below, most probably representing seedlings; trunk c.40 cm in diameter at base, decreasing to c.30 cm just below first dichotomy and decreasing at each successive dichotomy to c.20 cm below ultimate crown; dichotomies in well-grown individuals 4–5, giving 16–32 crowns, rarely 6 with 64 crowns, exceptionally branched once only or unbranched; trunk surface grey with close nodes c.2 cm apart, cracked vertically. Leaves long persisting, unless disturbed, as tangled spiny skirt covering trunk, usually burnt off leaving the trunk clean; ± 15 green living leaves in each crown; leaf base sheathing, split centrally to form a triangular cleft c.50 cm long; petiole from 80 cm long in exposed leaves to 1.25 m in juvenile plants, to 3.5 cm wide, semicircular in cross-section, yellowish in colour, except at edges where black, with a central black line below, armed with fierce distally pointing black spines to 2 cm long, 6 mm wide, much smaller in juvenile leaves; lamina strongly costapalmate, the costa sometimes twisted, to 80 cm long, with a spread of c.1.25 m, generally folded along costa, divided to third radius into c.30 segments on each side of costa, segments further divided a third; hastula black, irregular, rarely symmetrical, usually lopsided, 3–4 mm high, with a ragged crest of hairs; segments stiff, up to 4 cm wide below, tapering gradually to 2-pointed tips, glaucous green with yellowish ribs; dark green interleaflet filaments prominent; black scales abundant, scattered over lamina surface; fluffy pale brown scales present on ribs of young leaves. Male inflorescence arcuate to 1.5 m long, 5 × 3 cm in diameter at base; peduncular bracts c.25 cm long, covered in buff brown hairs and dark scales; partial inflorescences up to 10, each with 1–3(5) rachillae held horizontally; axis of first order branch c.25 × 1.5 cm; rachillae c.25 × 1 cm wide; bracts on rachillae with an exposed area c.4 × 1 mm, green-brown; pit hairs dull buff brown. Male flowers with pedicel c.1 mm long; sepals 1.5 × 0.5 mm; petals rounded, c.2 mm long, pale green; stamens bright canary yellow, filaments c.1.5 mm long, anthers rounded, c.0.5 mm long; no detectable scent. Female inflorescence as the male but rachillae usually solitary, sometimes paired, rarely in 3s, to 25 × 1.5 cm; pit c.10 × 8 mm at anthesis, densely filled with pale brown hairs; pedicel c.5 mm long, lengthening to over 10 mm at fruit maturity, c.6 mm in diameter, including densely packed hairs; sepals and petals c.5 × 7 mm wide, bluntly triangular, bright green; ovary c.5 mm wide, pinkish brown or

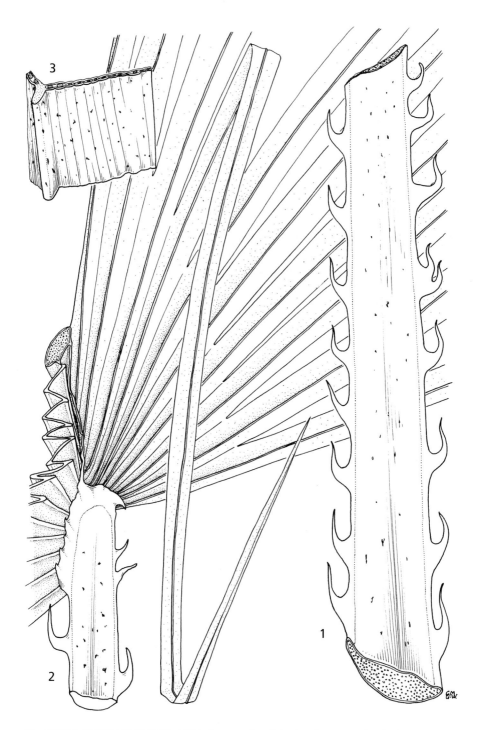

Fig. 13.2.**13**. HYPHAENE CORIACEA. 1, part of petiole showing upward pointing spines (× ²/₃); 2, base of leaf blade showing irregular hastula (× ²/₃); 3, part of leaf-segment showing dot-like scales (× 2). All from *Dransfield* 4817. Drawn by Christine Grey-Wilson. From Flora of Tropical East Africa.

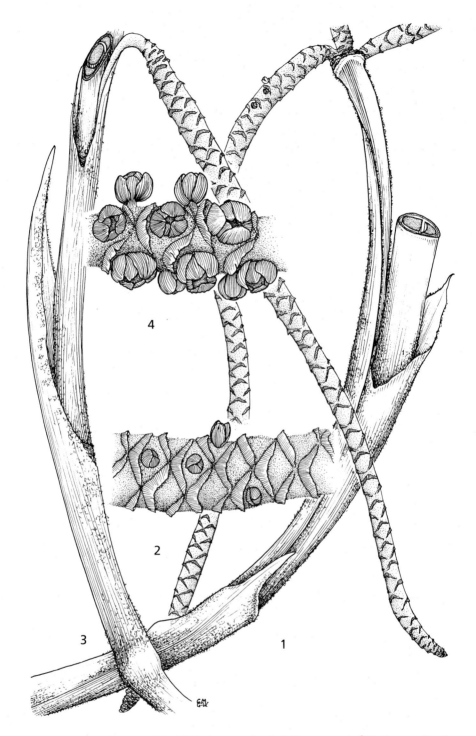

Fig. 13.2.**14**. HYPHAENE CORIACEA. 1, part of male inflorescence ($\times$ $^2/_3$); 2, part of male rachilla ($\times$ 2); 3, part of female inflorescence ($\times$ $^2/_3$); 4, part of female rachilla ($\times$ 2). 1–2 from *Dransfield* 4793, 3–4 from *Dransfield* 4817. Drawn by Christine Grey-Wilson. From Flora of Tropical East Africa.

pale straw-coloured; stigma nectariferous. Young fruit dull maroon-brown to straw-coloured with greenish-cream pits, often with waxy bloom. Mature fruit rich orange-brown to deep chestnut, rarely pale golden brown, extremely variable in shape and size from (6)7–10(12) × (4)5–8(9) cm wide at widest point, from almost cylindrical to oblong, obovoid, or sometimes much deformed, almost always with 2 distinctly compressed, flattened or sunken lateral faces; epicarp dimpled with conspicuous pock-marks, moderately shiny, flecked with sinuous pock markings, tough except on old rotted fruit; mesocarp pale golden brown, usually richly aromatic (like ginger-bread) in ripe fruit, to 1 cm or more thick; pyrene usually oblong to obovoid with 2 flattened faces; endocarp 4–8 mm thick at equator and c.10 mm thick at terminal pore; pore 10–13 mm in diameter; endosperm c.7 mm thick.

Mozambique. N: Palma Dist., between Quionga and Macanga beach, 10 m, 17.iv.1964, *Torre & Paiva* 12128 (K, LISC, SRGH). MS: Machaze Dist., Maringa, Save R. 180 m, fl. 21.vi.1950, *Chase* 29451 (K, SRGH). GI: Inhassoro Dist., Bazaruto Is, fl. x.1958, *Mogg* 28503 (K).

Along the Eastern African coast, extending from Kenya and Somalia south to Mozambique. Coastal lowlands and extending inland along watercourses, sometimes partly cultivated in cleared areas, its distribution much influenced by humans; 0–100 m.

Conservation notes: Widespread coastal species; not threatened.

As with *Hyphaene coriacea, H. compressa* has been referred to by many names and it has been difficult to interpret how the names have been used. It is a highly polymorphic taxon.

3. **Hyphaene coriacea** Gaertn. in De Fruct. **1**: 28, t.10 (1788). —Kirk in J. Linn. Soc. **9**: 234 (1866). —Furtado in Garcia de Orta **15**: 448 (1967). —Tuley in Palms Africa: 25 (1995). —M. Coates Palgrave, Trees Sthn. Africa, ed.3: 99 (2002). Types: Eastern Africa, uncertain locality (?Malindi), *Berkhey* s.n. (TUB holotype, FI). FIGURES 13.2.**13** & **14**.

    Hyphaene natalensis Kuntze in Linnaea **20**: 15 (1847). —Furtado in Gard. Bull., Sing. **25**: 283 (1970). —Drummond in Kirkia **10**: 232 (1975). Type: South Africa, Durban, 1843, *Geinzius* (not traced).

    Hyphaene turbinata H. Wendl. in Bot. Zeit. **39**: 92 (1881). —Wright in F.T.A. **8**: 123 (1901). Type: "Central Africa", no locality, ix.1858, *Kirk* s.n. (K holotype).

    Hyphaene coriacea var. *minor* Drude in Bot. Jahrb. **21**: 126 (1895). Type: South Africa, Pondoland, n.d., *Bachmann* (B† holotype).

    Hyphaene turbinata var. *ansata* Becc., Palme Borass.: 39 (1924). Type: Tanzania, Rovuma, n.d., *Busse* 1167 (B† holotype, FI).

    Hyphaene pyrifera Becc. var. *gosciaensis* (Becc.) Becc., Palme Borass.: 38 (1924). Type: Somalia, Goscia, Basso Giubo, fruit sent to Beccari, Inst. Colon. Italiano, no collector, no date (FI).

    Hyphaene pileata Becc., Palme Borass.: 42 (1924). Type: Tanzania, ?Rufiji Dist., Kiombani, n.d., *Zimmermann* III (FI holotype).

    Hyphaene tetragonoides Furtado in Garcia de Orta **15**: 459 (1967). Type: Mozambique, between Quisanga and Pemba (Porto Amelia), n.d, *Mendonça* s.n. (LISC holotype, COI, K, LMU).

Clustered or solitary palm, occasionally already mature while still a rosette, tending to form shrubby thickets with decumbent trunks, rarely more than 5 m tall, very rarely tree-like. Stems suckering, hence building up clumps of uneven aged trunks; dichotomizing once or twice, rarely more, when engulfed by sand dunes the dichotomies close, producing a clump of crowns on the dune surface; stems grey, to 25 cm in diameter, usually less, often covered by long-persistent leaf bases, which can be destroyed by fire leaving very close leaf scars. Crown consisting of 8–15 green leaves; petiole to 70 cm long, often much shorter, to 3 cm wide, widening at base, the triangular cleft c.20 cm long; petiole greyish-green, armed along its length with forward-pointing black triangular spines to 1 cm long, with rich brown scales and thin to very dense wax, especially when young; hastula regular or very asymmetrical. Lamina strongly costapalmate, the costa 30–80 cm long, usually curved, grey-green to very glaucous, with a dense covering of wax in young leaves (especially in sand dune forms) and scattered black scales; lamina divided into 15–20 segments, to $^1/_2$ radius in mid-leaf and $^3/_4$ radius near edge, segments to 3.5 cm broad, rather stiff, rarely exceeding 40 cm. Male inflorescence to 1 m, pendulous or arching, with 5–7 partial inflorescences; peduncle c.3 cm wide; peduncular bracts c.20 cm long, covered in abundant brown hairs and white wax; rachillae 1–3 in a group, slender, rarely exceeding 15 × 0.7 cm; rachilla-bracts 1 × 4 mm wide, exposed at anthesis. Male flowers with narrow sepals, c.3 × 0.5 mm wide; corolla stalk c.2 mm long, lobes somewhat spathulate, 3 × 1.5 mm wide; filaments c.2 mm long, anthers c.1.5 mm long, bright yellow. Female inflorescence as male but rachillae usually 1–2 only in each partial inflorescence; pits c.4 × 7 mm at anthesis. Female flower with pedicel c.3 × 1.5 mm wide at anthesis; sepals and petals similar, c.2 × 2 mm at base, triangular; ovary bright green, c.2 mm wide. Fruit extremely variable in size and shape, from ovoid and pyriform to strongly expanded distally, rarely 6 × 4 cm wide at apex, usually much less, occasionally as small as 3 × 2.5 cm, usually with a distinct ridge on one side and distinct widening toward apex; epicarp generally matt or very finely dimpled, without prominent warts or pockmarks, pale green when immature, ripening mid to dark brown; mesocarp c.4 mm thick, faintly aromatic, dark brown; endocarp c.3 mm thick except at apical pore. Seed ± polyhedral, conforming to shape of endocarp; endosperm c.5 mm thick.

Mozambique. N: Mogincual Dist, Mogincual near Quinga beach, 5 m, fl.& fr. 28.iii.1964, *Torre & Paiva* 11454 (K, LISC, SRGH). Z: Maganja da Costa Dist., 35 km on road from Maganja to Namacurra, 27.i.1966, *Torre & Correia* 14189 (K, LISC). GI: Massinga Dist., Massinga, Pomene, 10 km from hotel towards Rio dos Pedras, 16.vii.1981, *de Koning & Hiemstra* 9056 (K, LMU, SRGH).

Found in Eastern Africa from Somalia to Mozambique, South Africa (Limpopo, Mpumulanga, KwaZulu-Natal) and Madagascar. Commonest in coastal regions, especially in sand dunes and beside creeks behind mangroves, much rarer inland; in sand dunes it sometimes acts as a pioneer stabilizer; 0–250 m.

Conservation notes: Limited distribution in the Flora area but not threatened; Least Concern.

Considerable difficulties have been experienced in the interpretation of the usage of the name *H. coriacea* in the literature. Sometimes it has been used correctly while in other places it has been used in place of *H. compressa*. Furtado's analysis of the typification of *H. coriacea* seems convincing and is followed here, but several of his taxa should be included within *H. coriacea*.

6. **BORASSUS** L.

Borassus L., Sp. Pl.: 1187 (1753). —Beccari, Palme Borass.: 2 (1924). —Bayton in Kew Bull. **62**: 561–586 (2007). —Dransfield *et al.*, Genera Palmarum, ed.2: 329 (2008).

Solitary, robust to massive, dioecious tree palms, not dying after flowering; stems sometimes swollen, unbranched, or branching due to injury. Leaves massive, palmate or costapalmate, often long-persisting but finally falling to leave a clean trunk; leaf sheath short, not clearly distinct from petiole, splitting longitudinally to produce a central triangular cleft; petiole stout, usually armed with irregularly curved teeth along the margins, terminating in a conspicious adaxial hastula and a much smaller abaxial hastula; lamina divided into numerous induplicate, usually bifid segments. Inflorescences axillary, interfoliar, male markedly different from female. Male inflorescence with an elongate peduncle with prophyll and several bracts; partial inflorescences few to numerous, each in axil of a rachis-bract, branch-axis joined to main axis for some distance above insertion; branch axis with a basal bare portion terminating in (1)3 or more rachillae; rachillae massive, cylindrical, bearing connate imbricate bracts, also partly joined to the axis to form pits, each enclosing a cincinnus of flowers and closed by free tip of the bract; cincinnus with 3 or more flowers. Male flowers emerging singly from pits; sepals 3, joined into a shallowly or deeply lobed tube; corolla stalk-like at base, lobes 3, imbricate; stamens 6, with subulate filaments and erect anthers; pistillode minute. Female inflorescence simple or with 1–2 branches, clothed in large imbricate connate bracts. Female flowers massive, solitary, in axils of these bracts, each subtended by 2 bracteoles; sepals 3, imbricate; petals 3, similar to sepals; staminodes 6, forming a ring with minute anther rudiments; ovary globose, 3-locular, each locule with a single orthotropous ovule; stigmas 3, very short; septal nectaries conspicuous. Fruit massive, 1–3-seeded, borne within persistent perianth-segments; epicarp usually smooth or cracked vertically; mesocarp fibrous and pulpy. Seeds each enclosed in separate endocarp; endosperm homogeneous with a central hollow; embryo apical.

A genus of 5 species in Africa, Madagascar, India, Indochina, eastern monsoon area of Indonesia, New Guinea and perhaps N Australia. *B. flabellifer* L. is widespread in cultivation in SE Asia and Malesia.

Borassus aethiopum Mart., Hist. Nat. Palm. **3**, ed.1: 221(1838). —Beccari in Webbia **4**: 325 (1914). —White, For. Fl. N. Rhod.: 11 (1962). —Russell in F.W.T.A., ed.2 **3**: 168 (1968). —Tuley in Palms Africa: 20 (1995). —M. Coates Palgrave, Trees Sthn. Africa, ed.3: 99 (2002). Type: Ghana, *Thonning* 258 (holotype lost, see Hepper, W Afr. Herb. Isert & Thonn.: 154 (1976), Martius' description based on Thonning's *B. flabelliformis*); Ghana, Weija, 24.iv.1957 (♀ & ♂), *Tomlinson* s.n. (BH neotype, see Bayton 2007). FIGURE 13.2.**15**.

   Borassus flabellifer L. var. *aethiopum* (Mart.) Warb. in Engler, Pflanzenw. Ost-Afr. **B**: 20 & **C**: 130 (1895). —Wright in F.T.A. **8**: 117 (1901).

   Borassus aethiopum var. *bagamojensis* Becc. in Webbia **4**: 337 (1914). Type: Tanzania, Bagamoyo, n.d., *Zimmermann* s.n. (FI holotype).

Solitary palm with trunk ultimately reaching 20–30 m, swollen at base (85 cm diameter), 40–50 cm dbh, often with a pronounced swelling to 80 cm diameter c.10 m above base, at first covered with persistent leaf bases, these falling to leave a grey surface marked with annular leaf scars. Leaves c.40 in crown at maturity; leaf base c.90 cm long, split longitudinally down the mid-line, eventually also splitting opposite petiole; petiole to 325 × 15 cm above base, 3 cm thick, tapering to 7.5 × 3 cm thick near lamina insertion, rounded below, flattened or concave above, edged with shiny, black, irregularly erose, sharp protuberances to 1–2 cm high; upper hastula slightly asymmetrical, c.1 cm high; lower hastula inconspicuous, c.3 mm high; rib short, c.10 cm long, hence blade slightly costapalmate; blade radius c.180 cm from hastula to apex, shorter in outer regions; segments 60–80 or more, not all held in the same plane, stiff and only drooping at tips in age, equalling ± half radius in length, 5 cm or more wide, bifid apically for up to 15 cm, with numerous very fine longitudinal veins, distinct lateral veins and minute distant pale brown scales. Male inflorescence to c.1.5 m long, with 3–6 partial inflorescences; peduncle c.50 × 3 cm wide at base; bracts tubular, to 45 × 7 cm, coriaceous, finely striate, split to c.17 cm, lower few bracts empty, upper subtending branches terminating in (1)3 rachillae; rachillae up to 35 × 3 cm, brownish green, densely covered in ± 8 spirals of imbricate bracts, free portion of bracts c.10 mm × 5 mm; cincinni with c.9 flowers, emerging from pit one at a time. Male flowers subtended by 2 small bracteoles c.3 × 1 mm; calyx to 7 mm long, tubular, with 3 very shallow to deep lobes; receptacle to 7.5 × 1 mm, with 3 small rounded, somewhat hooded corolla lobes c.2 mm long; stamens 6; filaments very short, with anthers c.2 × 0.5 mm; pollen yellow. Female inflorescence 1–1.5 m long, with 2–3 empty bracts to 50 × 7 cm, occasionally branched once or twice; flower-bearing portion of axis 50–75 × 4–5 cm in diameter; bracts densely crowded, 3 × 12 cm wide, united laterally and forming pits. Female flowers solitary, with 2 bracteoles, each 2 × 3 cm wide; sepals 3, imbricate, 1.5 × 2 cm; petals 3, imbricate, 1 × 1.5 cm; staminodal ring with 6 teeth to 4 mm high; ovary c.1 cm in diameter, tipped with 3 stigmatic areas. Ripe fruit usually dull orange-brown, somewhat shiny, with much enlarged calyx and corolla; sepals in fruit 6 × 10 cm; petals in fruit 7–8 × 12 cm, erose-margined; fruit broadly ovoid, usually flat-topped, shape depending on number of seeds developing and on close-packing in infructescence, up to 12 × 14 cm wide or more, obscurely 3-angled; mesocarp to 1 cm thick, yellow and fragrant when ripe, numerous longitudinal fibres

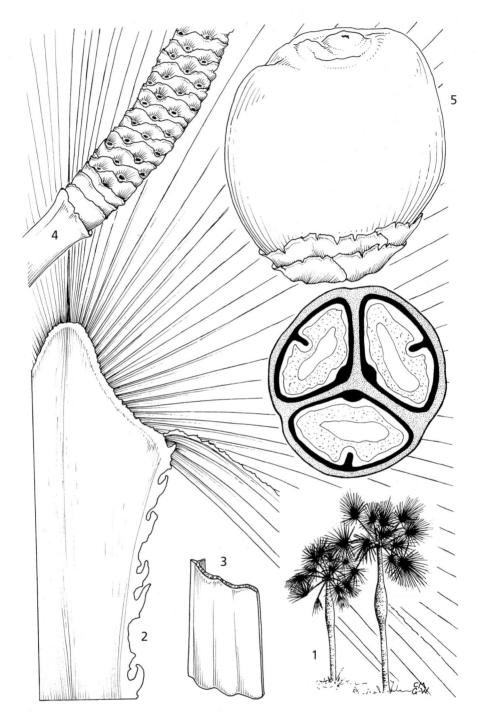

Fig. 13.2.**15**. BORASSUS AETHIOPUM. 1, habit ($\times \frac{1}{250}$); 2, base of leaf-blade showing hastula and petiole tip with irregular spines ($\times \frac{2}{3}$); 3, part of blade-segment ($\times$ 2); 4, part of male rachilla ($\times \frac{2}{3}$); 5, mature fruit ($\times \frac{1}{2}$); 6, transverse section of mature fruit ($\times \frac{1}{3}$). 1 from a photo of *Pawek* 13220, 2–3 from *Dransfield* 4811, 4 from *Bronn* 1904, 5 from *Dransfield* 4818, 6 from *Burtt* 1659. Drawn by Christine Grey-Wilson. From Flora of Tropical East Africa.

embedded in soft pulp; pyrenes up to 3, 10 × 10 × 5 cm, endocarp c.7 mm thick with 2 radially aligned internal longitudinal flanges, the outer to 1 cm high, inner to 5 mm, penetrating the seed. Seeds 1 in each pyrene, endosperm hard, bony white, homogeneous, ± completely filling pyrene cavity except for a small central space; embryo apical, top-shaped, c.5 × 4 mm.

Zambia. S: Mazabuka Dist., Mazabuka, Nampa Estates, 7.vii.1963, *van Rensburg* 1624 (K, SRGH). Also sight records from B:, N: (Luapula valley), C: (Lusaka) and E: (Luangwa valley) (M. Bingham, pers. comm.). **Zimbabwe**. E: Chimanimani Dist., Ngorima communal land, near Haroni–Rusitu confluence, 10.i.1969, *Biegel* 2782a (SRGH). S: Mwenezi Dist., Bangala Falls, Mtilikwe R., ♂ 12.xii.1953, *Wild* 44830 & 44909 (K, SRGH). **Malawi**. N: Rumphi Dist., 9.6 km W of Rumphi, 6.xi.1977, *Pawek* 13220 (K, SRGH). **Mozambique**. N: Nangade Dist, floodplain of Rio Metambue, between Pundanhar and Nagade (sight record, J. Timberlake, 2009). Z: Morrumbala Dist., crossroads to Morrumbala and Mopeia, 31.vii.1949, *Barbosa & Carvalho* 3799 (LMA). MS: Gorongosa Dist., Gorongosa Nat. Park (sight record, B. Wursten, 2006). GI: Inharrime Dist., Nhacongo, 13.vi.1981, *Marime & Manhiça* 58 (LMA).

Widespread throughout the less dry areas of tropical Africa down to N South Africa and N Madagascar but rarely in humid areas of W Africa. Occurring along water courses and in drier areas where there is a high watertable, often in magnificent dense stands; 0–400 m.

Conservation notes: Widespread and common in east of Flora area, although often used for construction; Least Concern. In Zimbabwe it is considered Critically Endangered (Sabonet Red Data List, Golding 2002).

The distribution of *Borassus* locally is very patchy, perhaps much influenced by humans and elephants. It has been suggested that it is naturalised rather than native in Zimbabwe (T. Müller, pers. comm.). J. Paiva (pers. comm.) records that elephant, man, and even lions are fond of the intensely fragrant fruit.

7. **COCOS** L.

Cocos L., Sp. Pl.: 1188 (1753). —Dransfield *et al.*, Genera Palmarum, ed.2: 416 (2008).

Tall, solitary, unarmed, monoecious tree palms, not dying after flowering; trunk with leaf-sheaths abscissing cleanly leaving leaf scars. Leaf base with continuous reticulate sheath, a triangular extension opposite the petiole. Leaves pinnate; leaflets regular, reduplicate, entire, single-fold. Inflorescence interfoliar, bisexual, branching to 1 order; prophyll inconspicuous, remaining between leaf-sheaths; peduncular bract 1, conspicuous, tubular at first, then splitting longitudinally, woody, longitudinally striate, beaked; peduncle moderately long; rachillae spreading, bearing 1 to few triads at base, solitary or paired flowers above. Male flowers with 3 imbricate sepals, 3 valvate petals and 6 stamens; pistillode small. Female flowers much larger than male, subglobose, with 3 large imbricate sepals and 3 large imbricate petals; staminodal ring inconspicuous; ovary large, ± spherical, tipped by a trifid sessile stigma. Fruit massive, with thick fibrous mesocarp, very hard endocarp with 3 basal pores, usually only 1 seed; endosperm hard, homogeneous, with central cavity partly filled with liquid; embryo basal. Seedling leaf simple.

One species of unknown origin, possibly W Pacific. Cultivated throughout the lowland tropics as a very important economic plant.

Cocos nucifera L., Sp. Pl.: 1188 (1753). —Kirk in J. Linn. Soc., Bot. **9**: 230 (1866). —Wright in F.T.A. **8**: 126 (1901). —Russell in F.W.T.A., ed.2 **3**: 161 (1968). —Tuley in Palms Africa: 100 (1995). Type: Tenga Rheede, Hort. Ind. Malab. **1**: 1–8, t.1–4 (1678–1703) [see Moore & Dransfield in Taxon **28**: 64, 1979]. FIGURE 13.2.**16**.

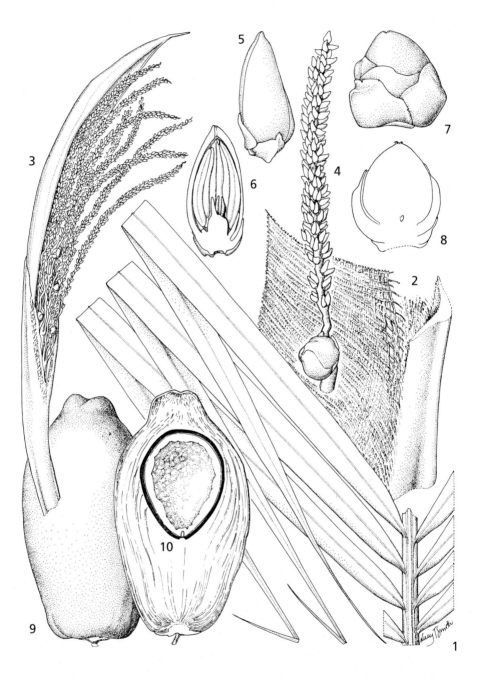

Fig 13.2.**16**. COCOS NUCIFERA. 1, mid-leaflets (× ¹/₄); 2, portion of leaf sheath (× ¹/₄); 3, inflorescence (× ¹/₄); 4, rachilla (× ¹/₂); 5, male flower (× 4); 6, male flower, longitudinal section (× 4); 7, female flower (× ³/₄); 8, female flower, longitudinal section (× ³/₄); 9, fruit (× ¹/₄); 10, fruitlets (× ¹/₄). All from a cultivated specimen at Kew 1969-4480. Drawn by Lucy Smith. From Baker & Dransfield, Field Guide Palms New Guinea (2006).

Palm varying from 'dwarfs' (trunks to 2 m only when first flowering) to tall forms with trunks to 30 m or more. Trunk usually somewhat swollen at base, to 80 cm in diameter, usually c.40 cm dbh, with conspicuous leaf scars 10–20 cm distant, commonly with pronounced vertical cracking of grey surface, especially at the base. Leaves up to 4–5 m long; petiole to 200 × 15 cm at base, channelled above, grey, scurfy hairy when young, bright pale green, golden yellow, or orangey in colour; leaflets single-fold, longest to 1 m long, to 2 cm wide, bright shiny green above, somewhat paler below, main vein usually prominent, pale green or golden yellow. Inflorescence interfoliar, to 1.5 m, usually less, bearing a prophyll c.50 × 12 cm wide, usually remaining between the leaf-sheaths; peduncular bract conspicuous, rather persistent, narrow, longitudinally striate, woody, brown, c.100 × 15 cm; peduncle to 40 cm long; rachillae ± 30 in all, cream-coloured, spreading, longest at the base, c.35 cm long, shortest the most distal, each with 1 to few basal triads, paired or solitary male flowers below, cream at anthesis. Male flowers somewhat asymmetric; sepals 3, imbricate, acute, 2 × 3.5 mm wide; petals 3, valvate, acute, flattish, fleshy, c.13 × 4 mm wide; stamens 6, erect filaments and narrow elongate anthers to 8 mm long, ± bifid apically; pistillode trifid. Female flower massive, much larger than male, to 3 cm in diameter; sepals 3, rounded, imbricate, c.1.5 cm wide; petals 3, rounded, c.2.5 cm wide; ovary rounded, to 2.5 cm in diameter. Fruit usually only 1 developing to maturity on each rachilla, massive, with basal persistent calyx and corolla increasing to 5 × 7 cm wide at maturity; whole fruit obovoid, obscurely 3-angled, extremely variable in shape and size, to 25 × 20 cm or more, usually only 1 of 3 carpels developing; mesocarp massive, fibrous; endocarp to 5 mm thick, extremely hard and woody, with 3 basal 'eyes', usually only one functional. Seed filling the large endocarp cavity, 10–15 cm in diameter; endosperm to 2 cm thick, lining the endocarp; embryo top-shaped, next to the functional 'eye'.

Mozambique. Known from sight records along the entire Indian Ocean coastline in N:, Z:, MS: and M:, although no herbarium specimens seen.

Widespread in coastal lowlands, performing best in more humid areas; also found inland as scattered planted individuals (said to be commonly planted in Zambia, M. Bingham, pers. comm.), usually below 400 m altitude, if above then not fruiting; 0–400 m.

Conservation notes: Widespread; not threatened.

For a discussion of the coconut as a plantation crop see Purseglove (Trop. Crops, Monocot. 2: 440–479, 1972). Harries (Bot. Rev. **44**: 265–319) gives an excellent review of the evolution, dissemination and classification of the coconut.

8. **ELAEIS** Jacq.

Elaeis Jacq., Select. Stirp. Amer. Hist.: 280 (1763). —Dransfield *et al.*, Genera Palmarum, ed.2: 451 (2008).

Solitary, robust, monoecious tree palms with pinnate leaves, not dying after flowering. Trunk erect or procumbent with tardily abscising leaf bases. Leaves large, with basal fibre spines, short acanthophylls developed from midribs of basalmost leaflets; leaflets numerous, inserted in one or more planes, reduplicate, with entire tips, single fold except at apex where sometimes compound (especially in juveniles), very rarely (as a monstrosity) all leaflets remaining joined together. Inflorescences usually male or female, sometimes mixed, branching to 1 order; prophyll and 1 large peduncular bract usually hidden among leaf-sheaths; peduncle rather short, bearing small, sometimes spine-like bracts; rachillae crowded together, densely floriferous except at spine-like tips. Male flowers solitary, partially sunken in pits, together with a membranous bracteole; sepals 3, free, imbricate; petals 3, free, imbricate or valvate; stamens 6, united basally by their fleshy filaments to form a staminal tube; pistillode with 3 small lobes. Female flowers much larger than male, usually with 2 abortive male flowers making a triad together with the 2 bracteoles; sepals 3, free, imbricate; petals 3, free, imbricate except for the triangular valvate tips; staminodal ring very small, 6-toothed; gynoecium with 3 conspicuous, fleshy, partially fused stigmas and a 3-locular ovary. Fruit with shining epicarp, oil-rich mesocarp and thick stony endocarp. Seed usually one only, covered in a thin integument; endosperm homogeneous, oil-rich, usually with a small central cavity; embryo apical next to one of the 3 endocarp pores ('eyes').

Two species – *E. guineensis,* the African Oil Palm, is found throughout the moister parts of Africa and E Madagascar and introduced into many tropical countries, and *E. oleifera* (Kunth) Cortes in eastern South America.

Elaeis guineensis Jacq., Select. Stirp. Amer. Hist.: 280, t.172 (1763). —Kirk in J. Linn. Soc., Bot. **9**: 231 (1866). —Wright in F.T.A. **8**: 125 (1901). —Russell in F.W.T.A., ed.2 **3**: 161 (1968). —Tuley in Palms Africa: 102 (1995). Type: Illustration t.172 in Select. Stirp. Amer. Hist. (1763). FIGURES 13.2.**17** & **18**.

Robust tree palm, in cultivated specimens often flowering while still trunkless. Trunk to 30 m tall, usually much less, 30–50 cm diameter, rarely broader, covered by remains of leaf bases when young, eventually becoming bare, but in high rainfall areas frequently obscured by epiphytes. Crown massive, consisting of 40–50 expanded leaves in a 8/13 phyllotaxis. Leaves to 7.5 m long in well-grown adults; leaf base long-persistent, with coarse brown fibres and upward pointing fibre-spines to 35 × 5 mm, c.10 mm distant, confined to sheath-margins; petiole to 1.25 m long, to 20 cm wide at base, armed below with bulbous-based spines to 4 × 1 cm wide, 1–5 cm distant, representing pulvini and midribs of basal-most leaflets; rachis semicircular in cross-section, tapering above, with 2 lateral grooves or faces; leaflets eventually 100–150 on each side, inserted rather irregularly in 2 planes, the whole leaf plumose, to 120 × 8 cm (in var. *idolatrica* A. Chev., a rare mutant, lamina remaining ± entire, not splitting into leaflets). Inflorescences either male or female, rarely bearing both and even more rarely with hermaphrodite flowers, individual palms passing through alternating phases of male and female inflorescence production; peduncle of male inflorescence 15–20 × 5 cm, densely hairy; rachillae crowded, ± 50 in number, 10–20 × 1–2 cm, with bare spine-like tip to 1 cm long; bracts to 3 × 1.5 mm. Male flower with 3 chaffy hooded sepals to 2 × 1.5 mm, and 3 chaffy acute petals to 2 × 1.5 mm; androecial tube to 2.5 mm long at anthesis; anthers reflexed, to 1.5 mm long. Female inflorescence more massive than male; rachillae much shorter, spine-like tip to 2 cm long. Female flower with 2 usually abortive male flowers and 2 bracteoles; sepals c.10 × 4 mm with rounded tips; petals similar to sepals; staminodal ring to 1 mm high, 6-toothed; ovary c.5 mm wide; style and stigma to 10 mm long. Fruit partially enclosed in enlarged calyx and corolla, tipped by stylar remains, very variable in size, 3–5.5 × 2–3 cm, ± asymmetrical, usually bright orange with dark red, almost black pigmentation in exposed upper parts; mesocarp 5–10 mm thick, yellowish, oil-rich; endocarp blackish brown, 2–5 mm thick. Seed usually 1 only, 2–3 × 1–1.5 cm, with thin integument; endosperm homogeneous with a narrow central cavity, oil-rich.

Malawi. N: Nkata Bay area (in White *et al.*, Evergr. For. Fl. Malawi: 107, 2001). **Mozambique**. N: Mecula Dist., along Rio Lugela downstream of Block C and Negamano (sight record J. Timberlake, 2003). MS: Marrromeu Dist., Zambezi delta, Coutada 11, 18°32.9'S 35°38.8'E, 28.vi.1999, *Timberlake & Müller* 4631 (SRGH).

Widespread throughout the tropics as a plantation oil-producing crop. In the Flora area confined to floodplain grasslands and near rivers; 0–150 m.

Conservation notes: Very localised and rare in the Flora area, where it is probably Vulnerable; not threatened globally.

It is not clear if the small populations found in the Flora area are naturalised, having arisen from individuals planted long ago, or truly native. However, the populations are certainly viable at these localities. In N Zambia (Mwinilunga, Luapula valley) it is said to be fairly widely planted (M. Bingham, pers. comm.), while in Zimbabwe a pilot plantation failed as the plants did not fruit owing to the cold.

The Oil Palm reaches its greatest abundance in SE Nigeria, lower Congo basin and moist areas of Sierra Leone and Benin. Small populations on the western coast of Madagascar were separated by Jumelle and Perrier de la Bâthie as var. *madagascariensis* Jum. & H. Perrier (in Palme Madag.: 78, 1913) and by Beccari as a

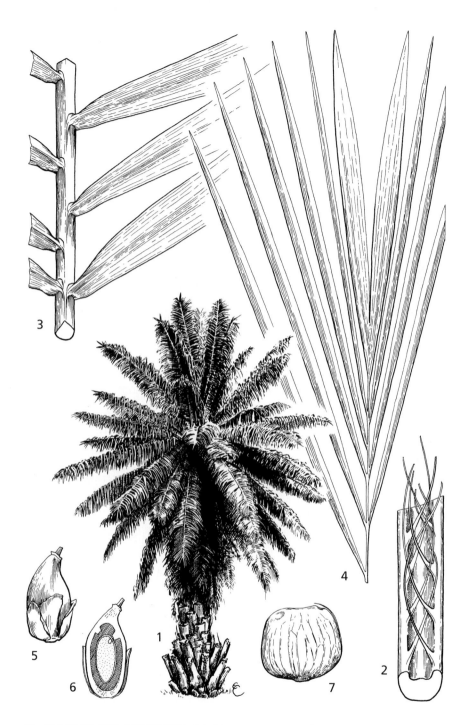

Fig. 13.2.**17**. ELAEIS GUINEENSIS. 1, habit of young plant (× ¹/₁₀₀); 2, portion of petiole with fibre-spines (× ²/₃); 3, mid-section of leaf rachis (× ²/₃); 4, leaf tip (× ¹/₄); 5, fruit (× ²/₃); 6, section of fruit (× ²/₃); 7, whole endocarp (× 1). 1 drawn from several photos, 2–4 from *Drummond & Hemsley* 3341, 5–6 from *Maggs* 1, 7 from sample cultivated in Sri Lanka. Drawn by Eleanor Catherine. From Flora of Tropical East Africa.

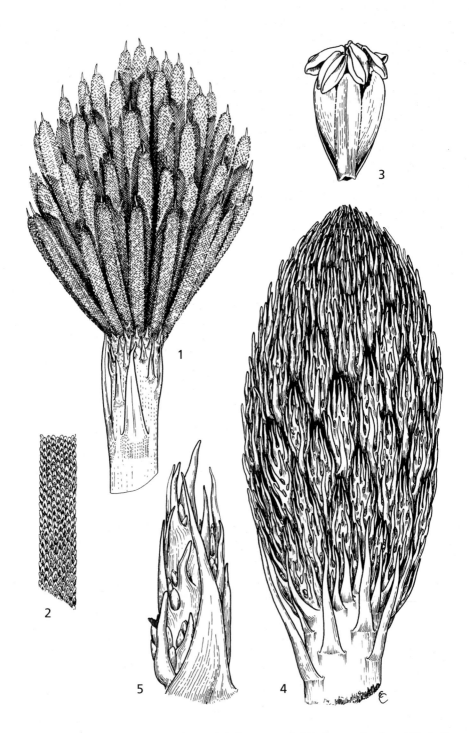

Fig. 13.2.**18**. ELAEIS GUINEENSIS. 1, male inflorescence ($\times \frac{1}{3}$); 2, part of male rachilla ($\times 1$); 3, male flower ($\times 10$); 4, young female inflorescence ($\times \frac{1}{3}$); 5, part of female rachilla ($\times 1$). From material cultivated in West Africa. Drawn by Eleanor Catherine. From Flora of Tropical East Africa.

distinct species *(E. madagascariensis* Becc. in Palme Madag.: 55, 1914), but because of the extraordinary variability of mainland *E. guineensis* the Madagascar plant fits within this range of variation.

Zeven (Semi-wild Oil Palm & Indust. Africa, Agric. Res. Rep. 689, Wageningen, 1967) and Hartley (The Oil Palm, 1988) have detailed discussions of the oil palm as a tropical crop.

PONTEDERIACEAE

by L. Catarino & E.S. Martins

Annual or perennial aquatic herbs, free-floating or rooted, with rhizomes or sympodial stems, rooting at nodes. Leaves sheathing at base, sometimes dimorphous – submerged leaves linear; emergent leaves with long, sometimes swollen petioles and orbicular, spathulate, ovate or lanceolate blades with curved nerves. Flowers in spikes, panicles or racemes, rarely single flowers, subtended by a spathe-like leaf sheath at the end of a 1-leafed section of sympodium. Flowers bisexual, nearly regular, occasionally cleistogamous. Perianth 6-merous, hypogynous, petaloid, segments united at base or free, equal or unequal, persistent. Stamens (1)3 or 6, equal or unequal, inserted on perianth, or hypogynous; filaments free; anthers with 2 chambers, opening by longitudinal slits or by terminal pores. Ovary superior (1)3-locular, with axile or parietal placentas, 1–many-ovulate; style filiform, entire or lobed. Fruit a capsule opening by 3 valves or an achene. Seeds small, ovate to subcylindrical or ellipsoid, longitudinally ribbed, with copious endosperm.

A family with 9 genera and c.35 species of freshwater aquatic plants from tropical and subtropical regions, mostly in South America. Five species native to Africa, with 4 in the Flora Zambesiaca area (see Catarino & Martins in Kirkia **16**: 21–26, 1996).

1. Stamens (1)3 . . . . **3. Heteranthera**
– Stamens 6 2
2. Perianth segments united below into a tube; flowers sessile or subsessile . . . . . .
. . . . **2. Eichhornia**
– Perianth segments free from base; flowers distinctly pedicellate . . **1. Monochoria**

1. **MONOCHORIA** C. Presl.

Monochoria C. Presl, Reliq. Haenk. **1**: 127 (1827).

Aquatic, glabrous, usually perennial herbs. Rhizome stout and creeping, or none. Leaves radical, often reduced to sheathing petioles and one in apex of each flowering stem (termed a bract by some authors). Flowering stems radical, each bearing one apical sheathing leaf (or bract), petiole appearing as a prolongation of stem. Inflorescence peduncle enclosed below in a membranous spathe, surrounded by sheath of apical leaf. Flowers with thin pedicels, arranged in shortly pedunculate, many-flowered racemes; perianth segments 6, free almost to base, subequal, blue or purple. Stamens 6, fixed near base of perianth segments, unequal, 5 short ones with yellow anthers, the other with a larger filament bearing an erect apical tooth and blue anther. Ovary 3-locular with axile placentas; ovules numerous; style filiform, stigma small. Capsule loculicidally splitting into 3 valves, with thin pericarp enclosed in persistent perianth; seeds numerous, nearly cylindrical, longitudinally ribbed.

A genus of c.8 species from the tropical Old World, extending from tropical Africa to Manchuria and S Australia. Only 2 species in tropical Africa – one in the Flora area, the other, *M. brevipetiolata* Verdc., from coastal West Africa.

Monochoria africana (Solms) N.E. Br. in F.T.A. **8**: 5 (1901). —Verdcourt in Kirkia **1**: 81 (1960); in F.T.E.A., Pontederiaceae: 3, fig.1 (1968). —Obermeyer in F.S.A. **4**(2): 63, fig.18 (1985). —Glen, Cook & Condy in Fl. Pl. Afr. **57**: t.2162 (2001). Type: Sudan, Equatoria, Jur Ghattas, 22.viii.1869, *Schweinfurth* 2296 (B† holotype, K, PRE). FIGURE 13.2.**19**.

   Monochoria vaginalis (Burm.f.) Kunth var. *africana* Solms in A. De Candolle, Monogr. Phan. **4**: 525 (1883). Type as above.

Aquatic herb, probably annual, up to 90 cm tall, without a rhizome; roots numerous, in a dense tuft. All or nearly all radical leaves with sheaths to 17 cm long. Flowering stems to 50 cm long. Apical leaves with petiole to 20 cm long, sheathing and enclosing spathe and base of inflorescence peduncle; lamina 8–12(14) × 5–10 cm, cordate to ovate, acute at apex, similar to one of radical leaves. Spathe to 3.5 cm long, membranous, with a subulate apex to 8 mm long. Raceme with a c.3 cm long peduncle; rachis 15–20 cm long; flowers 20–35, alternate to opposite; pedicels 6–10 mm long, slender. Perianth segments 6, free nearly to base, 12–17 mm long, the outer 3 c.3 mm broad, oblong, obtuse, the inner 3 5–6 mm broad, elliptic, obtuse, longitudinally nerved, blue to violet, with red punctuations longitudinally at mid-line. Stamens 6, dimorphic, 5 short ones 6–8 mm long with yellow anthers, a longer one 8–10 mm long with a lateral tooth near filament apex and bluish anther; anthers basifixed. Ovary superior, elliptic; style thin, 6–7 mm long; stigma papillate. Capsule 5–12 mm long, fusiform, enclosed in persisent perianth; persistent style forming a beak. Seeds c.0.7 × 0.5 mm, numerous, shortly cylindrical.

Malawi. S: Machinga Dist., Liwonde Nat. Park, near Park office, fl.& fr. 17.iv.1980, *Blackmore, Brummitt & Banda* 1247 (K). **Mozambique**. GI: Chicualacuala Dist., Vila Eduardo Mondlane (Malvérnia), fr. 25.iii.1974, *Balsinhas & Santos* 2684 (LMU).

Widespread in tropical and subtropical Africa, but known from relatively few and widely scattered collections at low altitude. In shallow pools, lagoons and on margins of rice fields; 450–600 m.

Conservation notes: Locally common at lower altitudes in the Flora area; probably Lower Risk near threatened. Regarded as Data Deficient in South Africa (Sabonet Red List database, Golding 2002).

The species is able to withstand great changes in water level. Juvenile plants are submerged, while adult plants are floating, emergent or terrestrial. It has in the past been mistakenly called *M. vaginalis*.

2. **EICHHORNIA** Kunth

Eichhornia Kunth, *Eichhornia* (1842); Enum. Pl. **4**: 129 (1843).

Annual or perennial aquatic herbs, free-floating or rooted and creeping. Sympodial stems rooting at nodes, with numerous long and feathery roots. Leaves alternate, occasionally in a rosette at base, sometimes dimorphous – submerged leaves linear, sheathing at base; emergent leaves with long, sometimes swollen petioles and orbicular, ovate, spathulate or lanceolate laminas. Flowers in spikes or panicles or single, surrounded at base by the petiole of a terminal leaf and a membranous bract (or spathe). Perianth funnel-shaped, blue or mauve. Stamens 6, the 3 upper ones included, 3 lower ones exserted; anthers dorsifixed. Ovary 3-locular; ovules numerous; style filiform. Capsule membranous, ovoid to fusiform, many-seeded, enclosed in pesistent perianth. Seeds small, subcylindrical to ellipsoid, finely ribbed.

Genus with 7–8 species from tropical America and Africa; some species widely introduced.

Leaves spaced on stem, dimorphous, submerged ones linear and sessile, floating ones with lamina ovate to subreniform; petiole long and weak; flowers single . . . .**1.** *natans*

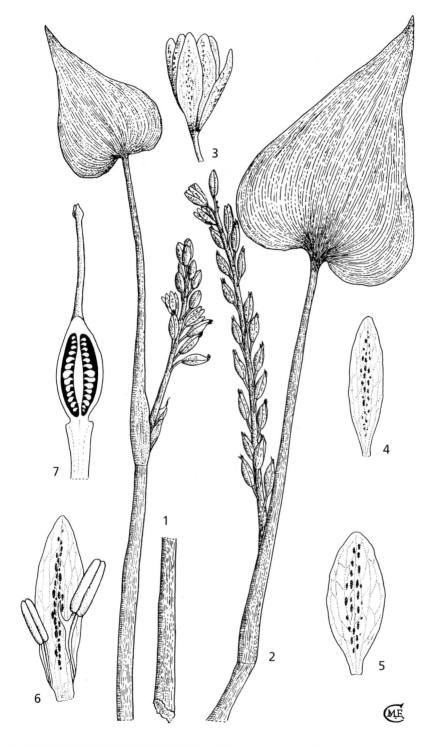

Fig. 13.2.**19**. MONOCHORIA AFRICANA. 1,2, flowering stems (× ²/₃); 3, flower (× 3); 4, outer perianth lobe (× 3); 5, inner perianth lobe (× 3); 6, perianth lobe with one each of two stamen types attached (× 4); 7, gynoecium in longitudinal section (× 6). All from *Greenway & Rawlins* 9483. Drawn by Maureen Church. From Flora of Tropical East Africa.

– Leaves arising at base of stem, uniform and all emergent; petiole long, stout and
usually swollen at middle; flowers in spikes . . . . **2.** *crassipes*

1. **Eichhornia natans** (P. Beauv.) Solms in Abh. Naturwiss. Vereine Bremen **7**: 254
(1882). —N.E. Brown in F.T.A. **8**: 4 (1901). —Verdcourt in F.T.E.A.,
Pontederiaceae: 6, fig.2 (1968). —Podlech in Merxmüller, Prodr. Fl. SW Afr.,
fam.154: 2 (1969). —Lawalrée & Bruynseels in F.A.C., Pontederiaceae: 4, fig.1
(1981). —Obermeyer in F.S.A. **4**(2): 64, fig.19 (1985). —Cook, Aq. Wetl. Pl.
Sthn. Africa: 232 (2004). Type: Nigeria, Oware, banks of Formosa R., *Palisot de
Beauvois* s.n. (G holotype). FIGURE 13.2.**20**.
   Pontederia natans P. Beauv., Fl. Owar. **2**: 18, t.68/2 (1810).

Aquatic herb, rooted, sometimes forming dense submerged mats in freshwater up to 2 m
deep. Submerged stems branched, up to 2 m long and 1.5–3 mm in diameter, rooting at lower
nodes; roots long, with numerous thin rootlets. Leaves alternate, dimorphic – submerged ones
30–70 × 1.5–4 mm, linear, sessile, sheathing at base, acute at apex; floating ones on upper
stem, petiole 25–65(85) mm long, lamina 8–25 × 6–22 mm, ovate to orbicular, cordate at base,
obtuse to subacute at apex, margins undulate. In shallow water only floating leaves may be
present. Flowers surrounded below by bract and petiole of the terminal floating leaf; peduncle
1.5–2.5 mm long; perianth purple, white or mauve, with cylindrical tube 15–18 × 2 mm and 6
subequal, obovate lobes c.7 × 2 mm; stamens free, unequal, filaments glabrous. Style long and
filiform. Capsule 10–20 × 1.5–3 mm, narrowly fusiform. Seeds numerous, c.1 × 0.4 mm,
longitudinally ribbed.

Caprivi Strip. Recorded as present in Clarke & Klaassen, Water Pl. Namibia: 80
(2001). **Botswana**. N: Okavango Swamps, Khwai R. upstream from Gobegha
Lagoon, 1000 m, fl. 7.iii.1972, *Gibbs Russell & Biegel* 1536 (COI, LISC, SRGH).
Zambia. B: Kaoma Dist., Luena R., Kaoma (Mankoya), fl. 20.xi.1959, *Drummond &
Cookson* 6677 (K). N: Luwingu Dist., Debenham channel, beyond Matongo, fl.
19.ii.1959, *Watmough* 277 (K, LISC, SRGH). S: Namwala Dist., Lochinvar Nat. Park,
fl.& fr. 13.iii.1972, *van Lavieren, Sayer & Rees* 737 (K). **Zimbabwe**. N: Binga Dist.,
Sebungwe, Chicomba Vlei, fl.& fr. 11.iii.1952, *Lovemore* 252 (K). S: Mwenezi Dist.,
Mwenezi R., gorge upstream from Buffalo Bend, fl.& fr. 28.iv.1961, *Drummond &
Rutherford-Smith* 7561 (LISC, SRGH).
   Widespread in tropical and subtropical Africa, but not in South Africa. Forming
dense, submerged, rooted mats in shallow (to 1 m deep), still or slow-flowing water
of rivers, lagoons and swamps, usually in full sun; 400–1400 m.
   Conservation notes: Widespread and common species; not threatened.
   Eichhornia natans is very similar to the tropical American species *E. diversifolia*
(Vahl) Urban. If they are found to be conspecific, the older name *E. diversifolia* will
take precedence.

2. **Eichhornia crassipes** (Mart.) Solms in A. De Candolle, Monogr. Phan. **4**: 527
(1883). —Wild in Kirkia **2**: 9 (1961). —Verdcourt in F.T.E.A., Pontederiaceae: 4
(1968). —Lawalrée & Bruynseels in F.A.C., Pontederiaceae: 5, fig.2 (1981). —
Obermeyer in F.S.A. **4**(2): 64 (1985). —Gopal, Water Hyacinth, Aq. Pl. Studies
1: 17–30 (1987). —Cook, Aq. Wetl. Pl. Sthn. Africa: 232 (2004). Type: Brazil,
Minas Gerais, *Martius* s.n. (M holotype).
   Pontederia crassipes Mart., Nov. Gen. et Sp. Pl. **1**: 9, t.4 (1824).

Herbaceous aquatic plants 20–80 cm tall, free-floating or rarely rooting on bottom. Roots
numerous, sometimes very long, densely feathery, apex with developed root cap. Stem short,
rhizomatous, with lateral stolons from which new plants originate (broken parts of stem also

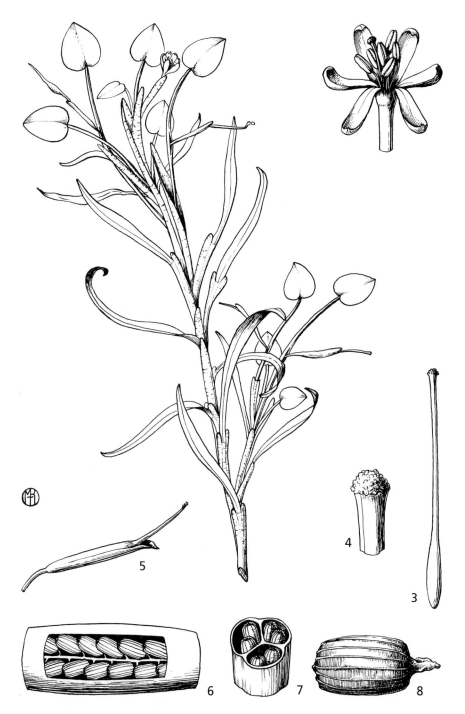

Fig. 13.2.**20**. EICHHORNIA NATANS. 1, flowering stem (× 1); 2, flower (× 4); 3, female flower parts (× 4); 4, tip of style and stigma (× 16); 5, capsule and part of spathe (× 2); 6, part of capsule, with wall partially removed to show seeds (× 8); 7, transverse section of capsule, oblique view (× 8); 8, seed (× 32). All from *Duke* 2. Drawn by Olive Milne-Redhead. From Flora of Tropical East Africa.

form new plants). Leaves 4–10, emerging from water, in a rosette, radical, erect; petiole 5–30(60) cm long, fusiform, swollen at middle or below, with well developed aerenchyma, enveloped at base by membranous stipules 2–15 cm long; blade up to 25 × 25 cm, usually smaller, largely ovate to orbicular, obtuse to rounded at base, obtuse at apex. Flowering stems with a small apical leaf, blade c.2.5 cm long, sheath c.6 cm long, enclosing spathe. Inflorescence a long, 4–15-flowered pedunculate spike, bending over after fertilization. Perianth tube 15–20 × 2–4 mm, cylindrical or slightly funnel-shaped, curved, pubescent; perianth lobes 6, white, blue or violet, elliptic, upper central one 30–35 × 15–20 mm, with yellow spot at centre, others 20–25 × 10–15 mm, without a spot. Stamens 6, 3 short, 3 long, with glandular-hairy filaments, curved upwards at apex, anthers c.2 × 0.5 mm, dorsifixed. Ovary ovoid; style shorter than longer stamens; stigma globose, glandular-lamellate. Capsule 3-locular; seeds numerous, usually apomictic, rarely produced and rarely germinating.

Zambia. S: Mazabuka Dist., Kafue Gorge below Kafue Dam, 950 m, fl. 6.i.1973, *Kornaś* 2929 (K). **Zimbabwe**. N: Lomagundi Dist., Manyame (Hunyani) R., 64 km from Chinhoyi (Sinoia), Silverside Farm, 1000 m, fl. 12.i.1960, *Phipps* 2491 (K). C: Harare Dist., near Harare, Lake Chivero (McIlwaine), st. 6.vii.1962, *Loveridge* 477 (K). S: Chiredzi Dist., Cheche R., Triangle Sugar Estate, st. 16.i.1976, *Poley* in GHS 242,231 (K, SRGH). **Malawi**. S: Nsanje Dist., Bangula lagoon, 60 m, fl. 22.iv.1980, *Blackmore, Dudley & Osborne* 1326 (K). **Mozambique**. Z: Mopeia Dist., Zambezi R., near Mopeia, fl.& fr. iv.1973, *Martins* s.n. (LMU). T: Cahora Bassa Dist., Zambezi R., c.10 km upstream of Cabora Bassa Dam, st. 31.x.1973, *Macêdo* 5341 (LISC, LMU). MS: Gorongosa Dist., Chitengo, c.10 km from Gorongosa Nat. Park camp, near old camp, fl. 2.v.1978, *Diniz* 154 (LMU). GI: Inharrine Dist., 60 km S of Inhambane, Ravene, st. viii.1956, *Gomes e Sousa* 1812 (COI, K). M: Manhica Dist., between Macia & Marracuene, fl. 29.xi.1941, *Torre* 3890 (C, EA, LD, LISC, M, WAG).

Native of South America, now widespread in many tropical and subtropical countries, although not recorded from Namibia (Clarke & Klaassen, Water Pl. Namibia: 80, 2001). In quiet or slow-moving freshwater and in nutrient-rich dams; 0–1000 m (to 1700 m in East Africa).

Conservation notes: Introduced species, now a noxious weed.

Introduced to southern Africa in 1884 (Cook 2004), Water Hyacinth is often dominant over large areas and has become a serious and noxious weed in many places, interfering with fishing and water transport. Juvenile plants are generally submerged, while adult plants are free-floating.

3. **HETERANTHERA** Ruiz & Pav.

Heteranthera Ruiz & Pav., Fl. Peruv. Prodr.: 9, t.2 (1794).

Perennial aquatic herbs; stems creeping and rooting at nodes. Leaves either all submerged, linear, or with ovate to reniform floating lamina, sessile or with long petioles, sheathing at base. Flowering stems bearing an apical leaf, the sheath of which encloses a membranous spathe. Flowers small, in spikes or solitary; occasionally cleistogamous flowers occur inside the spathe or mingled with open flowers. Perianth funnel- or salver-shaped, nearly regular, divided into 6 equal or subequal lobes. Stamens 3, exserted, unequal, inserted at throat of the perianth (some authors refer to a unique stamen in cleistogamous flowers). Ovary unilocular with 3 parietal placentas, or imperfectly 3-locular; ovules numerous; style filiform, stigma small. Capsule oblong or linear with numerous, ovoid, longitudinally ribbed seeds.

A genus of c.10 species, only 1 native to tropical Africa, the others in tropical and subtropical America.

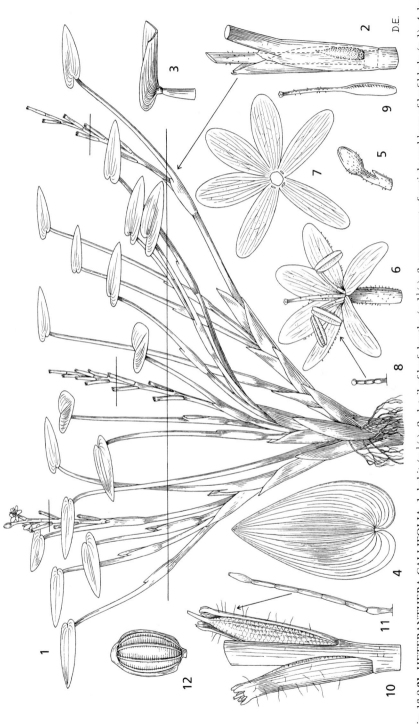

D.E.

Fig. 13.2.21. HETERANTHERA CALLIFOLIA. 1, habit (× 1/2); 2, detail of leaf-sheaths (× 1½); 3, upper part of petiole and base of leaf blade (× 1); 4, leaf blade, surface view (× 1); 5, bud (× 4); 6, flower (× 4); 7, perianth, detached and flattened (× 4); 8, glandular hair from perianth (× 40); gynoecium (× 4); 10, part of fruiting body with two capsules (× 2); 11, glandular hair from perianth remnants sheathing capsule (× 40); 12, seed (× 22). 1–4, 10–12 from *Polhill & Paulo* 1704, 5–9 from *Polhill & Paulo* 2168. Drawn by Derrick Erasmus. From Flora of Tropical East Africa.

Heteranthera callifolia Kunth, Enum. Pl. **4**: 121 (1843) as "*callaefolia*". —N.E. Brown in F.T.A. **8**: 2 (1901). —Verdcourt in F.T.E.A., Pontederiaceae: 6, fig.3 (1968). — Podlech in Merxmüller, Prodr. Fl. SW Afr., fam.154: 2 (1969). —Lawalrée & Bruynseels in F.A.C., Pontederiaceae: 8 (1981). —Obermeyer in F.S.A. **4**(2): 68, fig.21 (1985). Type: Senegal, *Sieber* 51 (P holotype, K, M). FIGURE 13.2.**21**.

Heteranthera kotschiana Solms in Schweinfurth, Beitr. Fl. Aethiop.: 205 (1867). —N.E. Brown in F.T.A. **8**: 3 (1901). Type: Sudan, Kordofan, by Mulbes at Obeid, *Cienkowsky* 378 (W holotype).

Monochoria vaginalis sensu Kirk in J. Linn. Soc., Bot. **8**: 147 (1865), non (Burm. f.) Kunth.

Aquatic herb up to 50 cm tall with creeping stem, rooting at lower nodes; roots densely covered with short rootlets. Leaves with ± erect petioles 6–15(20) cm long, sheathing at base, lamina floating to emergent, 2–7.5 × 1–5 cm, ovate to largely ovate, cordate at base, obtuse to subacute at apex. Flowers sessile, glandular hairy, up to 12(20) in a spike surrounded at base by a spathe and dilated petiole of a terminal leaf; rachis 5–10 cm long, just raised above water. Sometimes inflorescence reduced to a basal cleistogamous flower included in spathe. Perianth with a nearly cylindrical tube 5–12 mm long; lobes 6, narrowly ovate, 5–7 mm long, white, blue or purple. Stamens 3, exserted from tube; filaments short, c.2 mm long; anthers dorsifixed near base, yellow. Ovary ovoid; stigma small. Capsule trigonous-oblong, 8–12 mm long in open flowers, 12–18 mm long in cleistogamous flowers, many seeded, apiculate, enclosed by persistent perianth. Seeds ovoid to cylindrical, ribbed.

Botswana. N: Moremi Wildlife Reserve, fr. 14.iv.1974, *P.A. Smith* 884 (K). SE: Gaborone, Content Farm, 1050 m, fl.& fr. 10.iii.1978, *Hansen* 3366 (K). **Zambia**. B: Kazungula Dist., ii.1911, *Gairdner* 431 (K). N: Mporokoso Dist., N end of Mweru Wantipa (Mweru-Wa-Ntipa), near Selemani, 1020 m, fl.& fr. 17.iv.1961, *Phipps & Vesey-FitzGerald* 3273 (COI, LISC). C: Lusaka Dist., Lazy J Ranch, 20 km SE of Lusaka, fl. 25.ii.1995, *Bingham* 10408 (K). S: Choma Dist., 1050 m, fl.& fr. 9.iii.1955, *Robinson* 1125 (K, LISC). **Zimbabwe**. N: Binga Dist., Lundum, Chizarira Nat. Park (Game Res.), fl. 24.ii.1974, *Thompson* 932 (K, SRGH). W: Bulilimamangwe Dist., edge of Simukwe R., 8 km downstream from Mt Jim, fl.& fr. 11.iv.1974, *Ngoni* 375 (K, LISC, SRGH). C: Harare Dist., 6 mile spruit, granite pool, 1350 m, fl.& fr. 3.iv.1952, *Wild* 3798 (K, LISU). S: Mwenezi Dist., Urumbo Pan, between Fishans and Kapateni, fr. 25.iv.1962, *Drummond* 7712 (K, LISC). **Malawi**. N: Nyika Plateau, Mwanemba, 2400 m, fl.& fr. ii/iii.1903, *McClounie* 97 (K). S: Mangochi Dist., by Nkopola Lodge, 20 km NW of Mangochi, fl. 18.ii.1979, *Brummitt* 15404 (K). **Mozambique**. MS: Chemba Dist., opposite Ancueza (N'kuesa), N of Sena, 12.iv.1860, *Kirk* s.n. (K).

Widespread in subtropical and tropical Africa. Not recorded from Caprivi, but found in ephemeral pans elsewhere in NE Namibia (Clarke & Klaassen, Water Pl. Namibia: 80, 2001). In shallow, usually temporary, waters of swamps, pools, and pans; 450–2400 m.

Conservation notes: Widespread and locally common; not threatened.

BROMELIACEAE

by J.M. Lock & M.A. Diniz

Herbs, usually epiphytic or epilithic, sometimes xerophytic, usually rosette-forming, a few with a short stems, rarely climbing. Leaves spirally arranged in a rosette, with open sheaths; ligule absent. Inflorescence terminal, many-flowered, racemose, spicate or paniculate, with coloured bracts. Flowers usually hermaphrodite, actinomorphic. Perianth 6-merous, in 2 whorls. Outer whorl with 3 usually green ovate overlapping segments, free or joined at base. Inner whorl of 3 oblong overlapping segments, longer than members of outer whorl, free or joined at base. Stamens slender, in two whorls of 3, often attached to inner perianth whorl. Anthers sometimes attached to each other forming a ring, dehiscing by longitudinal slits, introrse. Style fleshy, free or joined to petaloid part of stamen and staminodes to form a tube; stigmas 3. Ovary inferior to superior, 3-locular; ovules few to numerous, on axile placentas. Fruit a berry or capsule, or rarely (*Ananas*) multiple and fleshy.

A family of about 46 genera and 2100 species, all from tropical America; one (*Pitcairnia feliciana* (A. Chev.) Harms & Mildbr.) from Africa (Republic of Guinea). The Pineapple (*Ananas comosus*) is cultivated throughout the tropics, often on a commercial scale. *Bilbergia pyramidalis* (Sims) Lindl. is reported as planted in Maputo, Mozambique (da Silva, Izidine & Amude, Prelim. Checklist Vasc. Pl. Mozamb. (2004).

ANANAS Mill.

Ananas Mill., Gard. Dict., abr. ed.4(1) (1754). —Smith & Downs in Fl. Neotrop. **14**: 2051 (1979).

Terrestrial rosette-forming herbs. Leaves linear, with hooked-spiny margins. Inflorescence shortly pedunculate, terminal, many-flowered, with a terminal rosette of leaves. Ovaries of all flowers fused to each other, to subtending bracts and to inflorescence axis, forming a large orange to yellow fleshy syncarp. Seeds usually absent in cultivated forms.

Genus of about 5 species, all native to E South America.

Ananas comosus (L.) Merr., Interpr. Herb. Amboin.: 133 (1917). Type in Herb. Amboinense (Phillipines).

    Bromelia comosa L. in Stickman, Herb. Amboin.: 21 (1754)
    Ananas sativus Schult. f. in Schultes & Schultes, Syst. Veg. **7**: 1283 (1830). Type from tropical America.

Terrestrial rosette-forming herb to 1 m. Leaves linear, to 100 × 5 cm, tapering to an acuminate apex, spiny on margins. Inflorescence terminal, many-flowered. Outer perianth segments blue. Fruit subcylindrical to broadly ovoid, to 50 × 30 cm, made up of fused inflorescence axis, bracts and ovaries of all flowers in the inflorescence, orange to yellow.

Mozambique: Recorded in da Silva, Izidine & Amude, Prelim. Checklist Vasc. Pl. Mozambique (2004) for Inhambane and Maputo Provinces. Widely planted, but no specimens seen.

A native of South America; perhaps originally from Paraguay but domesticated and spread in pre-Columbian times.

Conservation notes: Cultivated crop plant.

The first description with an illustration of the fruit reproduced widely was made by Fernandez de Oviedo y Valdez. Widely introduced to other areas by Portuguese and Spanish traders, reaching the Philippines by 1558 (where piña cloth, made from the white silky fibre extracted from the leaves, was first made about 1570), West Africa by 1602, and South Africa by 1660. Now pantropical and sometimes becoming weedy. Cultivated commercially in many countries, and traded as juice, as canned fruit, and as whole fresh fruit. Further details on economic and cultivation aspects can be found in J.L. Collins, The Pineapple (1960).

The edible variety is var. *comosus*, while var. *variegatus* (Lowe) Moldenke, with green, yellow and pink stripes is sometimes planted as an ornamental.

MAYACACEAE

by J. Cowley

Small perennial (sometimes annual?), glabrous, freshwater aquatic herbs, submerged or floating, occasionally creeping, rooting adventitiously at the base, usually branching sympodially above. Stems densely covered by spirally arranged 1-nerved, linear or filiform leaves, margins entire, apex usually bidentate, leaf sheaths absent. Flower bracts membranaceous, broadly ovate, shorter than leaves, soon splitting. Flowers bisexual, terminal, actinomorphic, solitary in leaf axils, 1–3 per branch, several clustered at apex of stems or forming a few-flowered umbel. Pedicels long, usually reflexed after flowering. Outer tepals 3, free, herbaceous, lanceolate, imbricate in bud, persistent. Inner tepals 3, shortly clawed, white, pink or violet, obovate or ± orbiculate. Stamens 3, free, opposite outer segments; filaments free, linear, short, slender; anthers 2–4-celled, erect, extrorse, basifixed, dehiscing by terminal pores, slits, short apical tubes or cups. Ovary sessile, globose, superior, unilocular, with 3 parietal placentae. Style filiform, elongate, persistent. Stigma entire, capitate or trifid. Ovules 6–30, biseriate. Fruit a 3-valved capsule. Seeds numerous, black, brown or red, ovoid or globose with basal hilum and beaked apex above the embryo; striate, rugulose, netted or pitted.

A family with 1 genus of c.4 species in warm temperate and tropical America and tropical Africa.

MAYACA Aubl.

Mayaca Aubl., Hist. Pl. Guiane **1**: 42 (1775). —Bentham & Hooker f., Gen. Pl. **3**(2): 843 (1883). —Faden in Fam. Monocot. (eds. Dahlgren, Clifford & Yeo), Mayacaceae: 387–388 (1985).

Description as for the family.

A genus comprising c.3 species in tropical America and 1 in tropical Africa, just entering the Flora area.

Mayaca baumii Gürke in Bot. Jahrb. Syst. **31** (69): 1 (1901). —N.E. Brown in F.T.A. **8**: 525 (1902). —Boutique in F.A.C., Mayacaceae: 2 (1971); in Distrib. Pl. Afr. **3**: map 90 (1971). Type: Angola, rio Quiriri, near Sakkamecho, 11.iv.1900, *Baum* 811 (?B holotype, G, K). FIGURE 13.2.**22**.

Stems to over 50 cm long, occasionally branched. Leaves dense, sessile, spirally arranged or subverticillate, filiform, linear-subulate, 10–17 × 0.7–1 mm, apex dentate. Inflorescence umbelliform, 1–3-flowered, near apex of stem. Floral bracts 3–4 × 1.5–2 mm, subtending a single flower. Pedicels 10–25 mm long, holding flowers above water, submerged when in fruit. Flowers white, outer segments lanceolate, persistent, obtuse, 5.5–7.5 × 1.5–2 mm; inner segments 6–10 mm. Filaments 0.6–1 mm; anthers 4-celled, widely cylindrical, cup-shaped, rounded at base, 1–1.3 × 0.5–0.8 mm, expanded rim uneven, opening internally below rim by an apical pore, covered by a convex operculum when immature. Ovary 1.2–1.6 mm long, containing 6 ovules. Style simple, 2–2.5 mm long, narrowing towards apex. Stigma entire, widening at slightly lobed, truncate apex. Capsules oblong-ovoid, 4–4.2 × 2.5–3 mm. Seeds 6, reticulately veined, globose, 1.5–1.8 mm wide, with a short beak.

Zambia. N: Samfya Dist., Lake Bangweulu, fl. 27.v.1964, *Fanshawe* 8689 (K, NDO, SRGH). W: Mwinilunga Dist., Chitunta R., 28 km N of Mwinilunga, fr. 10.vi.1963, *Drummond* 8276 (K, LISC, SRGH).

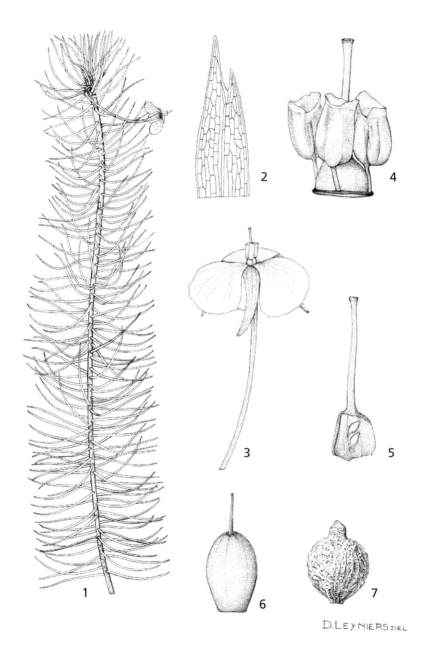

D.LEYNIERS DEL

Fig. 13.2.**22**. MAYACA BAUMII. 1, fragment of shoot (× 1); 2, top of leaf (× 100); 3, flower (× 3); 4, gynoecium and androecium (× 12); 5, gynoecium, partially cut longitudinally (× 12); 6, capsule (× 5); 7, seed (× 12). All from *Malaisse* 5204. Drawn by D. Leyniers. Reproduced with permission from Flore d'Afrique Centrale.

Also in S Congo and Angola. Freshwater aquatic of shallow lakes, swamps, pools, ponds or swiftly flowing streams, flowering and fruiting January to June; 1200–1400 m.

Conservation notes: Species only just enters the Flora area in N Zambia close to the Congo and Luapula river systems. Probably Lower Risk near threatened.

Plants found creeping above the waterline have a modified morphology compared to constantly submerged plants.

INDEX TO BOTANICAL NAMES

FAMILIES OF VASCULAR PLANTS REPRESENTED IN
THE FLORA ZAMBESIACA AREA

PTERIDOPHYTA
(Flora Zambesiaca families and family number. Published 1970)

Actiniopteridaceae		Gleicheniaceae	9	Parkeriaceae	
see Adiantaceae	18	Grammitidaceae	20	see Adiantaceae	18
Adiantaceae	18	Hymenophyllaceae	15	Polypodiaceae	21
Aspidiaceae	27	Isoetaceae	4	Psilotaceae	1
Aspleniaceae	23	Lindsaeaceae	19	Pteridaceae	
Athyriaceae	25	Lomariopsidaceae	26	see Adiantaceae	18
Azollaceae	13	Lycopodiaceae	2	Salviniaceae	12
Blechnaceae	28	Marattiaceae	7	Schizaeaceae	10
Cyatheaceae	14	Marsileaceae	11	Selaginellaceae	3
Davalliaceae	22	Oleandraceae		Thelypteridaceae	24
Dennstaedtiaceae	16	see Davalliaceae	22	Vittariaceae	17
Dryopteridaceae		Ophioglossaceae	6	Woodsiaceae	
see Aspidiaceae	27	Osmundaceae	8	see Athyriaceae	25
Equisetaceae	5				

GYMNOSPERMAE
(Flora Zambesiaca families and family number. Volume 1(1) 1960)

Cupressaceae	3	Cycadaceae	1	Podocarpaceae	2

ANGIOSPERMAE
(Flora Zambesiaca families, volume and part number and year of publication)

Acanthaceae	–	–	Balsaminaceae	2(1)	1963
Agapanthaceae	13(1)	2008	Barringtoniaceae	4	1978
Agavaceae	13(1)	2008	Basellaceae	9(1)	1988
Aizoaceae	4	1978	Begoniaceae	4	1978
Alangiaceae	4	1978	Behniaceae	13(1)	2008
Alismataceae	12(2)	2009	Berberidaceae	1(1)	1960
Alliaceae	13(1)	2008	Bignoniaceae	8(3)	1988
Aloaceae	12(3)	2001	Bixaceae	1(1)	1960
Amaranthaceae	9(1)	1988	Bombacaceae	1(2)	1961
Amaryllidaceae	13(1)	2008	Boraginaceae	7(4)	1990
Anacardiaceae	2(2)	1966	Brexiaceae	4	1978
Anisophylleaceae			Bromeliaceae	–	–
see Rhizophoraceae	4	1978	Buddlejaceae		
Annonaceae	1(1)	1960	see Loganiaceae	7(1)	1983
Anthericaceae	13(1)	2008	Burmanniaceae	12(2)	2009
Apocynaceae	7(2)	1985	Burseraceae	2(1)	1963
Aponogetonaceae	12(2)	2009	Buxaceae	9(3)	2006
Aquifoliaceae	2(2)	1966	Cabombaceae	1(1)	1960
Araceae	–	–	Cactaceae	4	1978
Araliaceae	4	1978	Caesalpinioideae		
Aristolochiaceae	9(2)	1997	see Leguminosae	3(2)	2006
Asclepiadaceae	–	–	Campanulaceae	7(1)	1983
Asparagaceae	13(1)	2008	Canellaceae	7(4)	1990
Asphodelaceae	12(3)	2001	Cannabaceae	9(6)	1991
Avicenniaceae	8(7)	2005	Cannaceae	–	–
Balanitaceae	2(1)	1963	Capparaceae	1(1)	1960
Balanophoraceae	9(3)	2006	Caricaceae	4	1978

Family	Vol(Part)	Year
Caryophyllaceae	1(2)	1961
Casuarinaceae	9(6)	1991
Cecropiaceae	9(6)	1991
Celastraceae	2(2)	1966
Ceratophyllaceae	9(6)	1991
Chenopodiaceae	9(1)	1988
Chrysobalanaceae	4	1978
Colchicaceae	12(2)	2009
Combretaceae	4	1978
Commelinaceae	–	–
Compositae		
tribes 1–5	6(1)	1992
Connaraceae	2(2)	1966
Convolvulaceae	8(1)	1987
Cornaceae	4	1978
Costaceae	–	–
Crassulaceae	7(1)	1983
Cruciferae	1(1)	1960
Cucurbitaceae	4	1978
Cuscutaceae	8(1)	1987
Cymodoceaceae	12(2)	2009
Cyperaceae	–	–
Dichapetalaceae	2(1)	1963
Dilleniaceae	1(1)	1960
Dioscoreaceae	12(2)	2009
Dipsacaceae	7(1)	1983
Dipterocarpaceae	1(2)	1961
Droseraceae	4	1978
Ebenaceae	7(1)	1983
Elatinaceae	1(2)	1961
Ericaceae	7(1)	1983
Eriocaulaceae	–	–
Eriospermaceae	–	–
Erythroxylaceae	2(1)	1963
Escalloniaceae	7(1)	1983
Euphorbiaceae	9(4)	1996
Euphorbiaceae	9(5)	2001
Flacourtiaceae	1(1)	1960
Flagellariaceae	–	–
Fumariaceae	1(1)	1960
Gentianaceae	7(4)	1990
Geraniaceae	2(1)	1963
Gesneriaceae	8(3)	1988
Gisekiaceae		
see Molluginaceae	4	1978
Goodeniaceae	7(1)	1983
Gramineae		
tribes 1–18	10(1)	1971
tribes 19–22	10(2)	1999
tribes 24–26	10(3)	1989
tribe 27	10(4)	2002
Guttiferae	1(2)	1961
Haloragaceae	4	1978
Hamamelidaceae	4	1978
Hemerocallidaceae	12(3)	2001
Hernandiaceae	9(2)	1997
Heteropyxidaceae	4	1978
Hyacinthaceae	–	–
Hydnoraceae	9(2)	1997
Hydrocharitaceae	12(2)	2009
Hydrophyllaceae	7(4)	1990
Hydrostachyaceae	9(2)	1997
Hypericaceae		
see Guttiferae	1(2)	1961
Hypoxidaceae	12(3)	2001
Icacinaceae	2(1)	1963
Illecebraceae	1(2)	1961
Iridaceae	12(4)	1993
Irvingiaceae	2(1)	1963
Ixonanthaceae	2(1)	1963
Juncaceae	–	–
Juncaginaceae	12(2)	2009
Labiatae		
see Lamiaceae, Verbenacaeae		
Lamiaceae		
Viticoideae,		
Pingoideae	8(7)	2005
Lamiaceae		
Scutellaroideae-		
Nepetoideae	–	–
Lauraceae	9(2)	1997
Lecythidaceae		
see Barringtoniaceae	4	1978
Leeaceae	2(2)	1966
Leguminosae,		
Caesalpinioideae	3(2)	2007
Mimosoideae	3(1)	1970
Papilionoideae	3(3)	2007
Papilionoideae	3(4)	–
Papilionoideae	3(5)	2001
Papilionoideae	3(6)	2000
Papilionoideae	3(7)	2002
Lemnaceae	–	–
Lentibulariaceae	8(3)	1988
Liliaceae sensu stricto	12(2)	2009
Limnocharitaceae	12(2)	2009
Linaceae	2(1)	1963
Lobeliaceae	7(1)	1983
Loganiaceae	7(1)	1983
Loranthaceae	9(3)	2006
Lythraceae	4	1978
Malpighiaceae	2(1)	1963
Malvaceae	1(2)	1961
Marantaceae	–	–
Mayacaceae	–	–
Melastomataceae	4	1978
Meliaceae	2(1)	1963
Melianthaceae	2(2)	1966
Menispermaceae	1(1)	1960
Menyanthaceae	7(4)	1990
Mesembryanthemaceae	4	1978
Mimosoideae		
see Leguminosae	3(1)	1970
Molluginaceae	4	1978
Monimiaceae	9(2)	1997
Montiniaceae	4	1978

Moraceae	9(6)	1991
Musaceae	–	–
Myristicaceae	9(2)	1997
Myricaceae	9(3)	2006
Myrothamnaceae	4	1978
Myrsinaceae	7(1)	1983
Myrtaceae	4	1978
Najadaceae	12(2)	2009
Nesogenaceae	8(7)	2005
Nyctaginaceae	9(1)	1988
Nymphaeaceae	1(1)	1960
Ochnaceae	2(1)	1963
Olacaceae	2(1)	1963
Oleaceae	7(1)	1983
Oliniaceae	4	1978
Onagraceae	4	1978
Opiliaceae	2(1)	1963
Orchidaceae	11(1)	1995
Orchidaceae	11(2)	1998
Orobanchaceae		
see Scrophulariaceae	8(2)	1990
Oxalidaceae	2(1)	1963
Palmae	–	–
Pandanaceae	12(2)	2009
Papaveraceae	1(1)	1960
Papilionoideae		
see Leguminosae	–	–
Passifloraceae	4	1978
Pedaliaceae	8(3)	1988
Periplocaceae		
see Asclepiadaceae	–	–
Philesiaceae		
see Behniaceae	13(1)	2008
Phormiaceae		
see Hemerocallidaceae	12(3)	2001
Phytolaccaceae	9(1)	1988
Piperaceae	9(2)	1997
Pittosporaceae	1(1)	1960
Plantaginaceae	9(1)	1988
Plumbaginaceae	7(1)	1983
Podostemaceae	9(2)	1997
Polygalaceae	1(1)	1960
Polygonaceae	9(3)	2006
Pontederiaceae	–	–
Portulacaceae	1(2)	1961
Potamogetonaceae	12(2)	2009
Primulaceae	7(1)	1983
Proteaceae	9(3)	2006
Ptaeroxylaceae	2(2)	1966
Rafflesiaceae	9(2)	1997
Ranunculaceae	1(1)	1960
Resedaceae	1(1)	1960
Restionaceae	–	–
Rhamnaceae	2(2)	1966
Rhizophoraceae	4	1978
Rosaceae	4	1978
Rubiaceae		
subfam. Rubioideae	5(1)	1989
tribe Vanguerieae	5(2)	1998
subfam.Cinchonoideae	5(3)	2003
Rutaceae	2(1)	1963
Salicaceae	9(6)	1991
Salvadoraceae	7(1)	1983
Santalaceae	9(3)	2006
Sapindaceae	2(2)	1966
Sapotaceae	7(1)	1983
Scrophulariaceae	8(2)	1990
Selaginaceae		
see Scrophulariaceae	8(2)	1990
Simaroubaceae	2(1)	1963
Smilacaceae	12(2)	2009
Solanaceae	8(4)	2005
Sonneratiaceae	4	1978
Sphenocleaceae	7(1)	1983
Sterculiaceae	1(2)	1961
Strelitziaceae	–	–
Taccaceae		
see Dioscoreaceae	12(2)	2009
Tecophilaeaceae	12(3)	2001
Tetragoniaceae	4	1978
Theaceae	1(2)	1961
Thymelaeaceae	9(3)	2006
Tiliaceae	2(1)	1963
Trapaceae	4	1978
Turneraceae	4	1978
Typhaceae	–	–
Ulmaceae	9(6)	1991
Umbelliferae	4	1978
Urticaceae	9(6)	1991
Vacciniaceae		
see Ericaceae	7(1)	1983
Vahliaceae	4	1978
Valerianaceae	7(1)	1983
Velloziaceae	12(2)	2009
Verbenaceae	8(7)	2005
Violaceae	1(1)	1960
Viscaceae	9(3)	2006
Vitaceae	2(2)	1966
Xyridaceae	–	–
Zannichelliaceae	12(2)	2009
Zingiberaceae	–	–
Zosteraceae	12(2)	2009
Zygophyllaceae	2(1)	1963